Computational Paleontology

Ashraf M.T. Elewa
Editor

Computational Paleontology

Springer

Editor
Prof.Dr. Ashraf M.T. Elewa
Minia University
Fac. Science
Dept. of Geology
Minia
Egypt
aelewa@link.net
ashrafelewa@ymail.com

ISBN 978-3-642-16270-1 e-ISBN 978-3-642-16271-8
DOI 10.1007/978-3-642-16271-8
Springer Heidelberg Dordrecht London New York

Library of Congress Control Number: 2011923531

Cover design: deblik, Berlin

Printed on acid-free paper

Springer is part of Springer Science+Business Media (www.springer.com)

Foreword

Computers and quantitative methods are fundamental tools in all branches of modern science, and paleontology is no exception. It has not always been this way, however. Quantitative approaches were of course always used by paleontologists, but the mainstream literature used to focus on qualitative description. In general, paleontology was surprisingly slow in adopting quantitative methods, compared with geology and particularly biology. One reason could be the idea that the fossil record is too incomplete for statistical treatment. What is the point of using sophisticated methods on such poor data? This is a misunderstanding – in fact the opposite is the case. It is precisely when the data are incomplete that we need the machinery of statistics to assess the effects of sampling. On the subject of mathematical modelling, a common objection is that the complexities of biological systems cannot be captured in a simple model. Again I would argue otherwise, that exactly when the system is complex beyond the capabilities of the human brain, a reduced model can lead to fundamental understanding by virtue of its very simplicity. After all, the purpose of modelling in paleontology is insight, not prediction.

A spectacular, early application of computers in paleontology was Raup's modelling of shell coiling. Another pioneer was Richard Reyment, who contributes to this volume. Now, computers are used almost everywhere in the paleontological work flow, from field work, data collection and visualization (Mallison; Poza-Rey; Stoinski; this volume) to morphometrics (Reyment) and data management (Skjerpen and Dolven). Quantitative methods are also fundamental in studies of paleoecology, development and evolution (Brusatte; Zachos and Sprinkle; Weaver; Petrakis). Paleontology as a science has improved as a consequence of this development. Quantitative approaches do not always give more "correct" answers, but they do make the arguments clearer and the results easier to falsify. Also, modern methods of data analysis and visualization have the power to suggest new research questions that would not have appeared otherwise.

There is something intriguing about the combination of modern technology and the vastness of geological time. The use of laser scanners, CT machines or DNA sequencing on fossils rarely fails to interest the media. This fascination was used to full effect in the blockbuster movie Jurassic Park (1993), where molecular biology and computer science interfinger with the horrors of the Mesozoic. This movie was also a technological breakthrough concerning the use of 3D computer graphics for visualizing ancient life forms. Such technology has since been used in countless movies and TV documentaries, contributing greatly to the present interest in paleontology among the general public.

Finally a piece of informal scientometrics: The ratio between hits for "computational paleontology" and "paleontology" on Google is presently 0.0036%. We therefore have some way to go compared with physics or biology, where the similar ratios are 1.3 and 2.0%. Slightly more alarming is the ratio for the "soft" science of archaeology, at 0.0049%. Hopefully, this book will contribute to us beating the archaeologists!

Oslo, Norway Oyvind Hammer

Contents

Contributors

Stephen L. Brusatte Division of Paleontology American Museum of Natural History, Central Park West at 79th Street, New York, NY, 10024, USA and Department of Earth and Environmental Sciences, Columbia University, New York, NY, USA, sbrusatte@amnh.org

Jane K. Dolven Department of Geosciences, University of Oslo, PO Box 1047, Blindern, Oslo, 0316, Norway, jane@radiolaria.org

Ashraf M.T. Elewa Geology Department, Faculty of Science, Minia University, El-Minia, Egypt, aelewa@link.net; ashrafelewa@ymail.com

Heinrich Mallison Museum für Naturkunde – Leibniz-Institut für Evolutions- und Biodiversitätsforschung an der Humboldt-Universität zu, Berlin, Germany

Panos V. Petrakis National Agriculture Research Foundation-NAGREF, Institute of Mediterranean Forest Ecosystem Research, Laboratory of Entomology, Terma Alkmanos, 115 28 Ilissia, Athens, Greece, pvpetrakis@fria.gr

Eva María Poza-Rey Centro Mixto UCM-ISCIII de Investigación sobre Evolución y Comportamiento Humanos, Instituto de Salud Carlos III, c/Monforte de Lemos, 3, 28029 Madrid, Spain, epoza@isciii.es

Richard A. Reyment Naturhistoriska Riksmuset, S 10405, Stockholm, Sweden, richard.reyment@nrm.se

Hans Skjerpen Virtual One AS, 0491, Oslo, Norway, hans@one.no

James Sprinkle Department of Geological Sciences, Jackson School of Geosciences, University of Texas, Austin, TX 78712-0254, USA epoza@isciii.es

Juan Luis Arsuaga Centro Mixto UCM-ISCIII de Investigación sobre Evolución y Comportamiento Humanos, Instituto de Salud Carlos III, c/Monforte de Lemos 3, 28029, Madrid, Spain and Departamento de Paleontología Facultad de Ciencias Geológicas, Universidad Complutense de Madrid, Ciudad Universitaria s/n, 28040, Madrid, Spain

Stefan Stoinski Computer Vision a Remote Sensing, Technical University Berlin, FR 3-1; Franklinstr. 28/29, D-10587 Berlin, Germany, stoinski@fpk.tu-berlin.de; http://www.cs.tu-berlin.de

Timothy D. Weaver Department of Anthropology, University of California, One Shields Avenue, Davis, CA 95616, USA, tdweaver@ucdavis.edu

Louis G. Zachos Nonvertebrate Paleontology Laboratory, Texas Natural Science Center (Texas Memorial Museum), 10100 Burnet Rd, Austin, TX 78758-4445, USA

Chapter 1
Computational Paleontology

Ashraf M.T. Elewa

Computational paleontology is simply a term applied to using computers and its facilities in the field of paleontology. However, we should be exactly specific in describing the term through explaining the main themes of this motivating and attractive scientific field.

Although the idea of using computer for solving paleontological problems is not new, but Oyvind Hammer, the famous Norwegian paleontologist and mathematician, introduced the term "computational paleontology" to the public through his computational paleontology webpage (1996). He described the term as the use of mathematical models, simulation, computer graphics and computers in paleontology. I know Oyvind since several years and I believe he is one of the pioneers in this field. He, together with David Harper and Paul Ryan, developed their Paleontological Statistics Software Package for Education and Data Analysis (PAST) in the year 2001 (see Hammer et al. 2001). Oyvind considered PAST as a follow-up to PALSTAT extensive package of Ryan et al. (1995).

It is worth mentioning that some paleontologists use the phrase "computer-assisted paleontology", some others prefer to use the idiom "computer-aided paleontology", still the expression "computational paleontology" sounds more relevant.

One of the earliest books to discuss the subject is that titled "Multidimensional Paleobiology" by Reyment (1991). Eight years later, Reyment and Savazzi (1999) introduced computational examples on the frequencies of fossils species as one of eight kinds of data encountered by geologists. They devoted chapter two of their book to describe the use of graphic software available on a CD accompanying the book.

Uhen (2000) mentioned the following criticism to the book titled "Numerical Palaeobiology: Computer-Based Modelling and Analysis of Fossils and Their Distributions", by Harper (1999): As for most people, much of what paleontologists do with computers is mundane and rather uninteresting. We type manuscripts,

A.M.T. Elewa
Department of Geology, Faculty of Science, Minia University, Minia, Egypt
e-mail: ashrafelewa@ymail.com, aelewa@link.net

A.M.T. Elewa (ed.), *Computational Paleontology*,
DOI 10.1007/978-3-642-16271-8_1, © Springer-Verlag Berlin Heidelberg 2011

we create figures, we send and receive e-mail. Dispensing with these sorts of applications, Harper presents an eclectic collection of papers on topics covering computer-based analyses of fossils, from individual specimens to the entire fossil biota. I was generally pleased with the individual chapters but was somewhat disappointed with the lack of discussion of the status of "numerical paleobiology" in general.

In a paper titled "Graphics in computational paleontology", which was published in the Computer Graphics and applications, IEEE, Figgins (2001) stated that we should search for the buried keys in the past to unlocking our understanding of the form. He added that we may still use tools such as picks and shovels to uncover them, but today paleontologists are also using computers and graphics to dig into the past.

David Lewin (2002) tried, by using sophisticated computer programs, to answer questions related to how fast the tyrannosaur can run and whether the triceratops sprawl or stand straight. His interest is to use combined techniques from computer-aided design, rapid prototyping, and biomechanics for developing more accurate theories of dinosaurs' posture and movements.

In August, 2005, Christoph P. Zollikofer and Marcia Ponce de Leon published a book titled "Virtual Reconstruction: A Primer in Computer-Assisted Paleontology and Biomedicine". They argued that virtual reconstruction serves as an introduction to the principles of three-dimensional visualization techniques as they relate to fossil reconstruction and reverse engineering.

One of the most effective ways to facilitate wide spreading of paleontology is electronic publications. Elewa (2007) discussed the efficiency of the electronic journal "Palaeontologia Electronica" in a paper titled "A Powerful Electronic Journal in the New Millennium".

Tammy Dunlavey, of the Department of Geology in the UB College of Arts and Sciences, and her colleagues tried to develop a computational method to morph fossils back to their original shapes by calculating and excising the deformation. Their goal was to develop computer programs that can reliably solve the deformation problems related to burial of fossils deep in layers of rocks for thousands or millions of years [see University At Buffalo (2004)].

There are several software packages that are used for mathematical calculations and graphical representations concerning paleontological data (e.g., PAST, IMP, TPS, MORPHEUS et al, MORPHOMATICA, ... etc.), however Tapanila (2007) produced a new software, which is a new Excel-based spreadsheet application of the Sepkoski Compendium designed for educational use in paleontology and historical geology courses.

In summary, it could be declared that almost multivariate geostatistics are not commonly observed as fascinating subject matter (see Reyment and Savazzi 1999), yet our area under contemplation "computational paleontology" is an exception.

Consequently, it is imperative to point out the main discussed topics in this volume in the following:

1. What is computational paleontology?
2. Computational taxonomy and systematics

3. Paleontological information systems (paleoinformatics)
4. Computational functional morphology
5. Computation of growth and form
6. Mathematics and statistics for paleontology
7. 2D and 3D graphical representations of paleontological data
8. Computational genetics and heritage
9. Future insights

Looking to the above mentioned topics, Elewa has published several papers on topics 2, 5 and 6 (e.g., Elewa and Ishizaki 1994; Elewa et al. 1995, 1999, 2001; Elewa 1997, 1998, 1999, 2002, 2003, 2004; Reyment and Elewa 2002; Elewa and Morsi 2004). Moreover, Elewa edited two books on the topic "moprhometrics" (2004, 2010), which is considered as one of the topics related to mathematical and graphical representations of forms; including fossils.

In an international Senckenberg conference and workshop titled "Paleontology in the 21st Century", Norman MacLeod and Robert Guralnick (1997, 2000) stated that paleoinformatics is that area of paleontology concerned the management of information, including the preservation of systematic information and expertise. They argued that because paleontology is such an information-rich and integrative field, the management of its data has always been problematic. I would add that another serious problem is located in the isolation between paleontologists and taxonomists in which each team is working without knowledge of the work of the other team. Therefore, it is compulsory to unify the nomenclatures of both teams under same identification, and then we can make accurate databases of taxonomic works, which will be applicable to the two teams.

In an interesting article related to computational functional morphology, Susan Rigby and Gavin Tabor (2006) used computational fluid dynamics in reconstructing the hydrodynamic properties of graptolites. They suggest that major improvements in our understanding of graptolite functional morphology will result from further use of this novel technique.

The uppermost aim of editing this book is to explain how computation could be competent in fetching fossils to life and the past to present. Computers for paleontologists save time and costs, interpret mysterious events precisely and accurately, visualize the ancient life definitely and undeniably.

Proudly, I could select an outstanding team of experts to write professionally on computational paleontology. No doubt, without their contributions this book could not be completed. I also would like to pass my great appreciation and respect to Oyvind Hammer for writing the forward to this book.

No doubt, computational paleontology techniques are frequently used by many students, researchers, and professionals. As well, this book introduces up to date information and useful ideas on the subject. I hope readers enjoy reading the chapters of this book in a manner promising to open a new gate to modern paleontology.

References

Elewa AMT (1997) Ostracode assemblages from the middle Eocene of the western bank of the Nile Valley between Samalut and Beni Mazar, Upper Egypt. N Jb Geol Paläont Abh, Stuttgart, 204(3): 353-378

Elewa AMT (1998) Fourier Biometrics: A case study on two species of the ostracode genus Bairdoppilata from the middle Eocene of Egypt. N Jb Geol Paläont Mh, Stuttgart, 1998(4): 203-211

Elewa AMT (1999) The use of allochthonus microfossils in determining palaeoenvironments: A case study using Middle Eocene ostracods from Wadi El Rayan, Fayoum, Egypt. Bull Fac Sci Assiut Univ, Egypt, 28(2): 33-52

Elewa AMT (2002) Paleobiogeography of Maastrichtian to Early Eocene Ostracoda of North and West Africa and the Middle East. Micropaleontology, USA. 48(4): 391-398

Elewa AMT (2003) Morphometric studies on three ostracod species of the genus *Digmocythere* Mandelstam from the middle Eocene of Egypt. Palaeontologia Electronica. 6(2): 11 pp

Elewa AMT (ed) (2004) Morphometrics – Applications in Biology and Paleontology. Springer-Verlag Publishers, Heidelberg, Germany

Elewa AMT (2007) A powerful Electronic Journal in the New Millennium. Palaeontologia Electronica, 10 (1; 2A): 2 pp

Elewa AMT (ed) (2010) Morphometrics for Non-morphometricians. Springer-Verlag Publishers, Heidelberg, Germany

Elewa AMT, Ishizaki K (1994) Ostracodes from Eocene rocks of the El Sheikh Fadl-Ras Gharib stretch, the Eastern Desert, Egypt (Biostratigraphy and paleoenvironments). Earth Science (Chikyu Kagaku), Tokyo, 48(2): 143-157

Elewa AMT, Morsi A A (2004) Palaeobiotope analysis and palaeoenvironmental reconstruction of the Paleocene-early Eocene ostracodes from east-central Sinai, Egypt. In AB Beaudoin, MJ Head (eds) The Palynology and Micropalaeontology of Boundaries. The Geological Society, London, 293-308

Elewa AMT, Ishizaki K, Nishi H (1995) Ostracoda from the El Sheikh Fadl-Ras Gharib stretch, the Eastern Desert, Egypt -With reference to distinguishing sedimentary environments-. In J Riha (ed) Ostracoda and Biostratigraphy. AA Balkema, Rotterdam, 203-213

Elewa AMT, Bassiouni MA, Luger P (1999) Multivariate data analysis as a tool for reconstructing paleoenvironments: The Maastrichtian to Early Eocene Ostracoda of southern Egypt. Bull Fac Sci Minia Univ, Egypt, 12(2): 1-20

Elewa AMT, Luger P, Bassiouni MA (2001) The Middle Eocene ostracods of Northern Somalia (Paleoenvironmental approach). Rev Micropaleont 44(4): 279-289

Figgins S (2001) Graphics in computational paleontology. Computer Graphics and applications, IEEE 21(6): 6-13

Hammer O, Harper D, Ryan P (2001) PAST: Paleontological Statistics Software Package for Education and Data Analysis. Palaeontologia Electronica 4(1): 9 pp

Harper D (ed) (1999) Numerical Palaeobiology: Computer-Based Modelling and Analysis of Fossils and Their Distributions. John Wiley and Sons, Chichester, England, 468 pp pp

Lewin DI (2002) Computer-Aided Paleontology: A New Look for Dinosaurs. Computing in Science and Engineering 4 (1): 5-9

MacLeod N, Guralnick R (1997) Paleoinformatics. Paleontology in the 21st Century, International Senckenberg Conference and Workshop, Frankfurt, Germany

MacLeod N, Guralnick R (2000) Paleoinformatics. In: Lane RH (sic), Steininger FF, Kaesler RL, Ziegler W, Lipps, J (eds). Fossils and the Future. Paleontology in the 21st Century. Senckenberg-Bücher Nr. 74: 31-37

Reyment R (1991) Multidimensional Paleobiology. Pergamon Press, Oxford, 377

Reyment RA, Elewa AMT (2002) Size and shape variation in Egyptian Eocene Loxoconcha (Ostracoda) studied by morphometric methods (a methodological study). In H Thiergärtner

(ed) Mathematical methods and data bank application in paleontology. Math Geol, Germany, 6: 3-14
Reyment R, Savazzi E (1999) Aspects of multivariate statistical analysis in geology. Elsevier 285 pp
Rigby S, Tabor G (2006) The use of computational fluid dynamics in reconstructing the hydrodynamic properties of graptolites. GFF, Sweden, 128 (2): 189-194
Ryan, P, Harper D, Whalley J (1995) PALSTAT, Statistics for palaeontologists. Chapman and Hall (now Kluwer Academic Publishers)
Tapanila L (2007) FossilPlot, an Excel-based Computer Application for Teaching Stratigraphic Paleontology Using the Sepkoski Compendium of Fossil Marine Genera. J Geoscience Education
Uhen M (2000) The Paleontological Legacy of Eckert and Mauchly. Paleobiology 26 (2): 310-315
University At Buffalo (2004, April 7) Paleontologists Use Computer To 'Morph' Deformed Fossils Back To Their Original Shapes. ScienceDaily. Retrieved from: http://www.sciencedaily.com/releases/2004/04/040407081228.htm; August 20, 2009
Zollikofer, De Leon MP (2005) Virtual Reconstruction: A Primer in Computer-Assisted Paleontology and Biomedicine. Wiley 333 pp

Chapter 2
Digitizing Methods for Paleontology: Applications, Benefits and Limitations

Heinrich Mallison

Introduction

Over the course of the last decades computers have evolved from a useful tool for rapidly calculating large amounts of equations to an indispensable part of everyday life. Today, cars will not run if a chip is faulty, and communication not only by phone and email, but also by conventional mail depends on computer codes. Computer generated or edited sounds and images dominate advertising, and their influence on education and especially entertainment is rapidly growing.

Slowly, computers have also found their way into science beyond the classic number-crunching applications in, e.g., climate modeling and statistical analysis. They can be useful tools for taxonomy, e.g., for cladistic analyses or for archiving and analyzing taxonomic data (see Elewa 2010). As pointed out by Elewa (2009), however, there is a bias in the use of modern computing techniques for vertebrate paleontology. At the other end of the scientific process, publication has also "gone digital", with journals such as PLoS One. The first fully online journal with a focus on paleontology is *Palaeontologia Electronia*, which started publishing online in 1998, open to any new and promising technique and data format for presenting scientific data and results (MacLeod and Patterson 1998). Authors are actively encouraged to experiment with new data formats. As pointed out by Elewa (2007), many research institutions sadly are slow in adapting to this novel way of knowledge distribution, and need convincing to accept e-publication as equal to conventional paper journals.

Because even detailed 3D objects can now be depicted and animated quickly on ordinary office and houschold computers, due to the high demand of many computer simulation games on 3D graphics power, an ever growing number of scientists use these powers for vertebrate paleontology. A classic method for obtaining 3D data on a fossil, prepared or unprepared, is via tomography. This is "the representation of three-dimensional structure as a series of two-dimensional images formed

H. Mallison
Museum für Naturkunde – Leibniz-Institut für Evolutions- und Biodiversitätsforschung an der Humboldt-Universität zu Berlin, Berlin, Germany
e-mail: heinrich.mallison@gmail.com

A.M.T. Elewa (ed.), *Computational Paleontology*,
DOI 10.1007/978-3-642-16271-8_2,

from parallel section" (Sutton 2008: p. 1587). Tomographic methods allow study both of the external shape and internal structures. Sutton (2008) details the various different techniques, as well as typical examples of their use in paleontology, so that only a short summary is required here.

The slices (or tomographs) of a tomography can either be studied directly, or used to create 3D visualizations of the fossil. They can be obtained using a variety of methods, some of which are destructive. These include the first use of tomographic study of fossils by Sollas (1904), who described a machine "which was designed to grind parallel sections of fossils at exact intervals of 0.5 mm or more" (Sollas 1904: p. 259). These sections were recorded by photography, and structures were hand-traced on the photographs. Sollas mainly applied his method to vertebrates (e.g., Sollas and Sollas 1913), other researchers adapted and used it on a wide variety of fossils (e.g., Simpson 1933; Ager 1965; Kermack 1970; Kielan-Jaworowska et al. 1986; Sutton et al. 2001; Bednarz and McIlroy 2009). The study of internal structures of brachiopods became highly reliant on serial grinding after Muir-Wood (1934) first applied this destructive technique. From such photographs 3D visualizations can be created even from data decades old, as long as the serial images show consistently interpretable color or structure differences between materials, and the intervals between them are known.

With the advent of modern scanning techniques non-destructive tomography became possible. Initially, many research approaches centered on X-ray Computed Tomography (CT, described in, e.g., Hounsfield 1980; Kak and Slaney 2001), scanning specimens to make internal structures visible. While individual tomograms can be used the same way as a physical section, the extraction and visualization of internal 3D shapes is of especially interest. One of the first fossils to be studied through X-ray CT scanning was the Eichstätt specimen of the famous *Archaeopteryx lithographica* von Meyer (1861) (Haubitz et al. 1988). This study used each slice singly, and was limited to fairly rough images. The first *Archaeopteryx* body fossil ever found, the London specimen, was later also scanned to gain information on the brain and inner ear (Alonso et al. 2004), which due to technical progress could be gleaned from high resolution 3D visualizations of the skull, both entire and in parts.

Today, because of the increasing availability of CT scanning opportunities (every hospital has at least one scanner), the use of CT scanning has practically become standard for specimens from which interesting discoveries are expected. Scans can also be used to plan, improve or replace preparation (e.g., Clark and Morrison 1994; Fraser et al. 2007), in which case the removal of artifacts is especially important (McLean et al. 2001), or even make preparation redundant. Still, the main use of CT data remains the study of already prepared fossils (e.g., Brochu 2000, 2003; Wedel 2003a, b; Sampson and Witmer 2007; Witmer and Ridgely 2008), used for research tasks that would be more difficult, more time consuming, or simply impossible if the real fossils were used.

There are several other non-destructive methods besides classic CT for obtaining tomogram stacks, including neutron tomography (NT, Schwarz et al. 2005) and, less suited for paleontology, magnetic resonance imaging (MRI, e.g., Clark et al. 2004). Most important, however, was the development of X-ray micro-tomography or high-resolution X-ray computed tomography (XMT, HRXT or HR-CT, e.g., Rowe et al. 2001; Carlson et al. 2003). This technique allows resolutions down to 5 μ, compared to several hundreds of microns for the best medical CT scanners. However, medical scanners were developed and optimized for the study of humans (Hounsfield 1980), and thus machines that allow scanning of specimens up to the size of obese human adults can be found in practically any large hospital. In contrast, micro-CT scanning is limited to much smaller objects, e.g., fossil fruits and seeds (DeVore et al. 2006), or small fossils preserved in amber (Dierick et al. 2007).

The University of Texas' Jackson School of Geosciences owns a HR CT scanner, which is used to supply the university's DigiMorph project, a special web library of CT-based 2D and 3D visualizations of the external and especially internal structures of extant and extinct vertebrates (digimorph.org), housing over a terabyte of data.

Synchotron radiation X-ray tomographic microscopy (SRXTM) uses a subatomic particle accelerator called a synchotron to create even higher resolutions and extremely crisp images, and was applied by Donoghue et al. (2006) to study fossil embryos shedding light on the initial radiation of multi-cell life. It has also been found by Smith et al. (2009) to be an excellent tool for virtual dissection of fossil and extant plants.

There are also a variety of methods to obtain digital representations of physical objects that scan only the external surfaces. Data can be gathered by (stereo-) photogrammerty, laser scanning, ultrasound or mechanical digitizing. Even caliper measurements can be translated into a 3D shape, if a sufficient number of them per surface area are taken in a defined reference system. The data type collected by laser, mechanical and sonic scanning is at first a series of point vectors. Various ways of transcribing this information to a computer aided design (CAD) program or other software exists, e.g., as individual point objects, as point clouds (groups of point objects), as curves (series of connected point objects) or as interpolated curves (mathematical formulae derived from a series of points). Often, polygon meshes are calculated from the point clouds. A polygon mesh, or polymesh, consists of a large number of triangles (and sometimes squares) connecting neighboring points, and is an easily transferable and editable way to represent a surface or body. The surfaces extracted from CT data are usually are also created as polymeshes.

Mechanical digitizing (Wilhite 2002; Mallison et al. 2009) involves a robotic arm, the tip of which is placed where one desires to record data, so that the density of surface point data collection is to a certain degree at the operator's discretion. This allows much smaller file sizes for simple fossil shapes than other digitizing methods, without extensive post-digitizing editing.

There are a variety of optical digitizing systems, only a small selection of which can be discussed here. For small specimens like mammal teeth, a number of techniques have been successfully employed, e.g., in for creating a paleoanthropological database (Kullmer et al. 2002), for studies on Recent and fossil human skull shapes (Subsol et al. 2002) and for studies on functional morphology of, e.g., the skull and teeth of bears (Engels 2007). Optical digitizing works through the principle of optic triangulation, using CCD cameras (CCD = charge coupled device; well known from digital cameras). The technique and mathematical methods involved are described in, e.g., Breuckmann (1993).

These and other digitizing methods each have their advantages and draw-backs, their cost and the time one needs to invest to learn and apply them differs greatly, and the potential uses of the resulting data, which depends mostly on the resolution and file size, also vary. Here, some important methods suitable for paleontology will be discussed in general terms (i.e., specific equipment or programs are not mentioned), to give a broad overview that helps researchers choose the method that best fits their intended research. Chapter 8 discusses photogrammetry and especially laser scanning in detail, thus these receive an abbreviated treatment here.

Institutional abbreviations:
GPIT – IFGT collection numbers
IFGT – Institute for Geosciences, Eberhard-Karls-University Tübingen
MB.R.### – Collection numbers of the MFN
MFN – Museum für Naturkunde – Leibniz Institute for Research on Evolution and Biodiversity at the Humboldt University Berlin

Applications, Benefits and Limitations

Digital files can take the place of the real physical object for many tasks, provided the accuracy is sufficient for the task at hand. For example, digitized ammonites can be measured digitally the same way real fossils can be measured with a caliper. However, in this case the gain from the digital file use is minimal or nonexistent. Worse, inaccuracies in the digital file stemming from careless digitizing may degrade the accuracy of the measurements, lowering the quality of any work based on the data. Digitizing a fossil just because it can be digitized is usually a waste of time, effort, and money. At the beginning of every project involving digitizing a detailed assessment of the digitizing requirements must be conducted, based on the intended use of the resulting files, with the specifications of the digitizing technique for accuracy, time required, and cost taken into account. Thus, before a discussion of the technical details of various digitizing methods for paleontology a few basic questions need answering: Why should anyone digitize a fossil? What can we use digital files for? What can be digitized, and which objects are not suitable? What are the advantages of using digital files for specific tasks, and what are the limitations that must be kept in mind?

Research

The most obvious use of a digital file is scientific research. The simplest form is the direct scanning of complete specimens as mentioned above for *Archaeopteryx*. Another typical example is the use of digital files of vertebrate bones in biomechanical studies (Stevens and Parrish 1999, 2005a, b; Hutchinson and Garcia 2002;

Hutchinson et al. 2005; Christian and Dzemski 2007; Mallison 2007, 2010a, b, c; Sellers et al. 2009). Real fossil bones of large animals are simply too heavy to handle easily, some (sauropod longbones and vertebrae) requiring two or more people to lift. If a task needs only the general shape of the bone, not internal structure, high resolution surface texture and color, it can be performed using CAD software and digital files. This saves time and also protects the fossils, because they have to be handled only once, for digitizing. For example, the digital files of GPIT/RE/7288 (GPIT 1 in the literature), a nearly complete individual of the prosauropod dinosaur *Plateosaurus engelhardti* von Meyer 1837 are available for researchers worldwide for study. I created and used them initially to create a virtual mount of *Plateosaurus* (Mallison 2007, 2010a, b). The next step involved "playing" with the articulated bones to determine the motion range of all joints (Mallison 2010b). This task was greatly aided by the fact that I did not have to create supports for the 239 separate bone files, but could manipulate them on the screen in any way I wanted. A major advantage of digital files is the ability to simply save any configuration at any time, and to compare different configurations directly next to each other. Today, scientists all over the world need only write an email to the Institute for Geosciences Tübingen curator, and can obtain the digital bone models for their own research via download.

Digital data in the form of complete vertebrate skeletons, either as complete scans of mounted specimens or as digital skeletal mounts, can be used for estimates of the volume, total mass, mass distribution and center of mass position of extinct animals. For example, Gunga et al. (2007, 2008) used scans of complete mounts of the sauropodomorph dinosaurs *Plateosaurus* and *Giraffatitan brancai* (Janensch 1914) (see Chap. 8) to create 3D volumes representing the reconstructed shape of the living animals, from which all the listed parameters could be calculated. In my own research I used the same model creation tools not on a scan of the existing skeletal mount of *Plateosaurus*, but on the virtual skeleton I built (Mallison 2010a, b in press). A less comfortable but also significantly smaller data base could also have been employed, consisting of artificial bone files created by (e.g., mechanically) digitizing one element of each type (one dorsal vertebra, one anterior dorsal rib, one posterior dorsal rib, one caudal vertebra, one tibia, etc.) and scaling copies to measurements of the other bones. Provided the elements of each group of bones used have sufficiently similar shapes, this is a "cheap&dirty" way to obtain a quite reliable representation of a large and multi-part fossil for relatively little digitizing time.

Digital files can furthermore be used for all kinds of shape analyses, including full surface representations or just selected data points for morphological landmark studies (e.g., Lohmann 1983; Bonnan 2004; Brombin et al. 2009; Bushman and Cornell 2009), using for example Procrustes analyses (Bookstein 1991; Mardia and Dryden 1998; Gower and Dijksterhuis 2004). Such analyses are helpful in paleontology, but also play an important role in medicine (e.g., DeQuardo 1999; Krey and Dannhauer 2008). Interestingly, if the intended landmarks are theoretically well-defined, they are easy to pick out for humans, so that different operators can gather data for one study (Brief et al. 2006). This could allow combining data from collections world-wide by email, without requiring one person to undertake expensive travels. Complete surfaces can be compared using techniques similar to those employed to test the accuracy of digitized data in (Mallison et al. 2009), which creates maps of shape differences.

An important advantage limited to the rock penetrating tomography techniques is the ability to retain the spatial associations between fossil parts that are normally lost during preparation (Sutton 2008). This can be especially important in connection with attempts to determine taphonomic deformation or puzzle together badly fragmented specimens. Furthermore, penetrating techniques allow the study of internal structures, easily the greatest contribution to paleontology by any digitizing technique. There are by now countless examples, including all kinds of fossils (Luo and Ketten 1991; Brochu 2000, 2003; Wedel 2003a, b; Alonso et al. 2004; Schwarz et al. 2005; Falk et al. 2007; Sampson and Witmer 2007; Witmer and Ridgely 2008; Balanoff et al. 2010; Fink and Humphries 2010).

One of the most difficult tasks in paleontology is the retrodeformation of fossil shapes, the attempt to reconstruct how a fossil was shaped before taphonomic or sedimentological processes, or worst of all metamorphosis, altered it, resulting in the loss of biological information (Hughes and Jell 1992). If the damage is solely brittle, resulting in breakage into pieces, the fossil must be puzzled together. This process can be time consuming and cumbersome, especially when parts are missing, and it is usually questionable whether there really is solely brittle deformation (Boyd and Motani 2008). Much worse is plastic deformation that alters the shape of the fossil, or a combination of both breakage and deformation. The topic is of extreme importance for comparative studies of systematics, phylogenetics and morphology (Angielczyk and Sheets 2007; Boyd and Motani 2008), and various techniques have been suggested for retrodeformation (e.g., Wood et al. 2003; Srivastava and Shah 2006; Shah and Srivastava 2007), including digital methods for 3D "jigsaw puzzling" (Zollikofer et al. 1995, 2005). For example, Ediacaran fossils were all soft-bodies, and depending on their orientation during flattening can be preserved as distinct shapes (Bamforth 2008: fig. 1-3.1), or alternatively appear similar as fossil when the living organisms differed (Bamforth et al. 2008). Additionally, tectonic deformation has significantly altered the fossils (Seilacher 1999; Wood et al. 2003; Ichaso et al. 2007), so that they now differ even more from their original size, orientation and shape. Both mathematical and photographic retrodeformation have been conducted (e.g., Rushton and Smith 1993; Seilacher 1999; Gehling et al. 2000; Wood et al. 2003; Bamforth et al. 2008); mathematical details can be found in (Wood et al. 2003). Sadly, many techniques do not work well (Angielczyk and Sheets 2007; Boyd and Motani 2008).

Taking the process of editing files a step further, instead of just removing deformation, missing parts can be replaced by scaled copies from other specimens, and small damaged areas digitally repaired. Extreme caution is required, because any repaired or composite digital file is speculative, and any science conducted using it is therefore at a risk of being less accurate than it would be if complete real specimens were used. However, digital repair can deliver best approximations that, cautiously used, may be much more useful than broken and incomplete specimens. Similarly, mirror images can be created with a single command, allowing easier comparison with other taxa if only contralateral elements are known. Additionally, digital data can be scaled at will, either allometrically or isometrically, which can make shape analyses easier.

Even reconstructions of soft tissues can benefit from digital data, e.g., the musculature of extinct vertebrates. Firstly, a digital model can be created, which, being 3D, can be much more accurate and informative than the classic 2D drawings, in which muscles are represented by lines. Cases in point are the highly details musculoskeletal models used by Hutchinson and Garcia (2002) and Hutchinson et al. (2005) of the hindlimb musculature of the theropod dinosaur *Tyrannosaurus rex* (Osborn 1905), and of another theropod, *Velociraptor mongoliensis* (Osborn 1924) as well as an Asian elephant by Hutchinson et al. (2008). Secondly, surface marks such as muscle and tendon attachment sites on bones are sometimes visible as rugosities, but most have no special texture. Almost all, however, lead to a slight flattening of the bone surface, which can be easily missed due to the colors and textures on a real bone, but are immediately visible on a shallow-angle oblique view of a (texture-free) digital file (Fig. 2.1).

In addition to body fossils, other structures of interest to paleontologists can be digitized as well. Some ichnofossils have complex 3D shapes that are difficult to

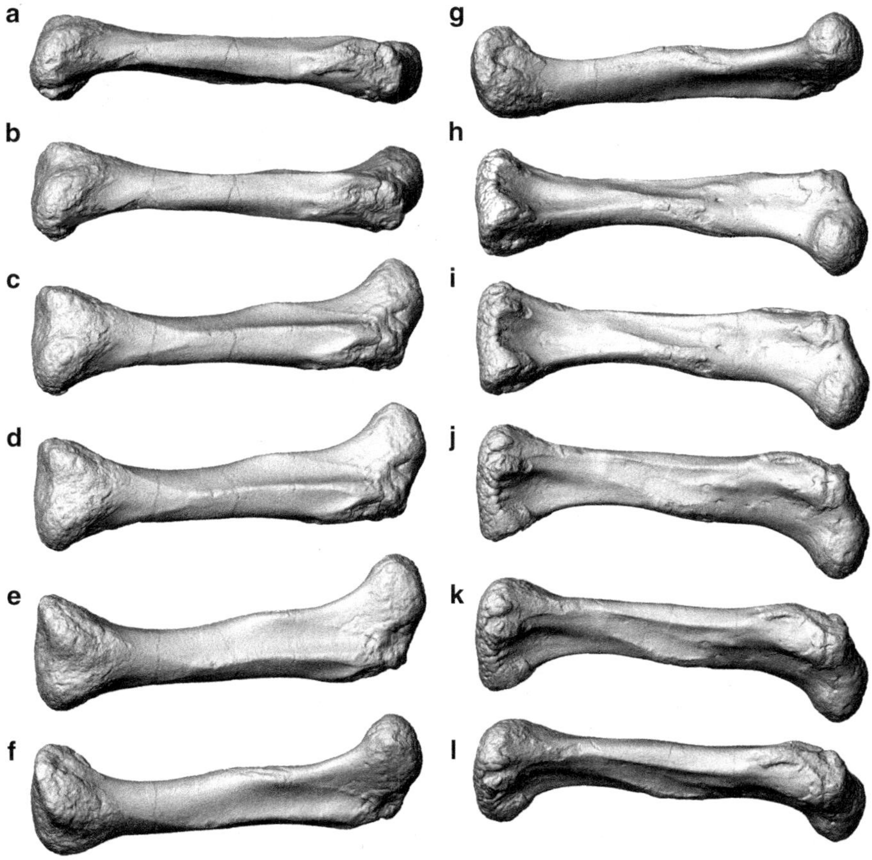

Fig. 2.1 High resolution laser scan based digital file of the left femur of *Kentrosaurus aethiopicus* (MB.R.4800.36). View rotated from (**a**) lateral, (**e**) anterior, (**g**) medial, (**i**) posterior and (**l**) slightly posterior of lateral. Note how ridges and rugosities are well visible at shallow angles

infer from the outside appearance. Bednarz and McIlroy (2009) created a 3D representation of a phycosiphoniform trace fossil through tomography, which sadly required the destruction of the specimen. However, the model allowed new insights into the complex 3D architecture of the ichnofossil. Similarly, digitizing of ichnofossils greatly facilitates the study not only of their shapes, but also of the process of their creation (e.g., Bates et al. 2009a; Raichlen et al. 2010). Also, predictions of footprint shape can be made not simply based on a rough assumption of the posture of a fossil foot and the soft tissues on it, but from complex motion cycle simulations and detailed soft tissue models (Mallison and Porchetti unpubl. data). A simplified version of such a predicted track is shown in Fig. 2.2.

These are but a few examples of past and present uses of digitizing in paleontology. They barely give an overview, but they highlight the impact of the digital age on the science.

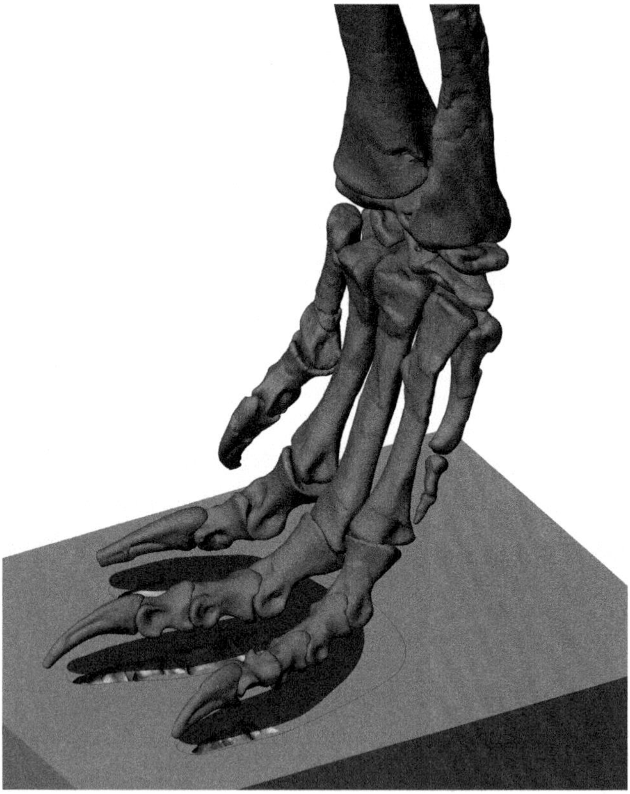

Fig. 2.2 The left foot of *Plateosaurus engelhardti* (part of GPIT Skeleton 2), digital CT-scan based files, is used to predict footprint shape. Here, solely the mid-stance position and the resulting print are shown. The *black curve* and the *dark shade* show the influence of two different soft tissue reconstructions (not shown) on the potential track. *Plateosaurus* left a tridactyle print when walking slowly; only at very high impact velocities caused by high speeds did the first toe possible touch the ground

Curation

Curators of paleontological collections must balance two main and often contrary aspects of their jobs: on one hand, the safekeeping of their precious specimens, on the other the accessibility that makes the collection scientifically useful in the first place. Additionally, some specimens usually are exhibited, both to provide justification to the public for the expense of collecting and preserving the material, and for educational purposes. Conflicts are pre-programmed, because access and exhibition result in handling of specimens, as well as exposure to dust, UV light, humidity and other potential dangers. All these problems can be totally or partly ameliorated by digitizing specimens.

Conservation and Accessibility for Research

Many vertebrate fossils, especially large bones, are rather robust and stand the occasional use by researchers without taking any damage. The other extreme are physically or chemically unstable fossils, such as a crinoid colony from the *Posidonia* Shale (Lias ε) of Dotternhausen, Germany, in the "Werksmuseum" of the Holcim Zement company that I had the pleasure to help prepare. The colony features some 20 short-stemmed crinoids on a 2 m long fossilized log. The sea lilies were mainly preserved as a mixture of pyrite and marcasite, two forms of iron sulfite (FeS_2). Marcasite is prone to react with moisture in the air and decay, the so-called "pyrite decay" (Rixon 1976; Newman 1998). Worse, sulfuric acid is produced in the chemical reaction, which spreads the destruction to other parts of the fossil and specimen labels (Stooshnov and Buttler 2001). The crinoid colony was thus potentially chemically unstable (sometimes, for unknown reasons, marcasite does not decay), and had to be kept at low relative humidity (Newman 1998) at all costs, otherwise it would crumble to dust over time. An extreme example of a decaying fossil is the early thyreophoran dinosaur *Emausaurus ernsti* (Hauboldt 1990), also from the Lower Jurassic of Germany. It has already been partly destroyed by chemical reactions caused by the wrong conservation methods being used on it, and by the decay of marcasite. It is unclear whether a re-preservation recently completed will stop the destruction. The mentioned crinoid colony is not of special scientific importance and solid enough to be casted. The *Emausaurus* fossils, in contrast, are brittle and represent the type and only specimen of a basal member of the armored dinosaurs, and they include a nearly complete skull. *Emausaurus* is therefore a prime candidate for modern touch-free digitizing methods. CT scanning can reveal both the external form and internal structures. High resolution laser scanning also preserves the external form, adding data on the coloration of the surfaces (if the scanner includes a color camera). This data can stand instead of the real fossils for many research tasks, and can be used to create 3D prints. Combining two instances of (nearly) touch free digitizing can thus preserve the exact external and internal shape, the color and some, but not all information on the details of surface structure.

Emausaurus was in fact CT scanned, for exactly these reasons, and as an aid for the re-conservation efforts (Hinz-Schallreuter pers. comm. 03/2010)

Aside from preserving fossils that decay on their own accord, digitizing can also protect other fragile fossils from destruction. A curator's and preparator's worst nightmare is a clumsy paleontologist let loose in the collection, breaking fossils by carelessly handling or even dropping them. But even the most careful researcher will occasionally damage a specimen, and some fossils are so fragile because of their delicate structure that even the most cautious handling will lead to destruction. If some research is conducted on digital files instead of the real specimens, the number of instances when the fossils must be handled can be reduced.

Furthermore, because CT scanning can reveal internal structures, it can not only be helpful for future preparation, but can aid in determining how fossils were treated during past preparation. Glued in metal armatures, as were the method of choice for mounting large skeletons, wires glued into the marrow cavity, plaster infills, etc. will all be easily visible on the scan slices, and can be created as separate 3D bodies during surface extraction, to be used as visual aids during re-conservation (Figs. 2.3 and 2.4).

Another important advantage of digital data compared to any physical object is that it can be multiplied flawlessly for an unlimited time, and can thus never be lost if sufficient precautions are taken. In contrast, even the sturdiest rock will be eroded over time, although it may take the touch of thousands of researchers to even produce a noticeable change to its external shape and appearance. Additionally, multiple copies mean that several researchers in different locations can study the same specimens at the same time, either isolated or as a cooperative effort, or work on material that is at the same time exhibited elsewhere.

Finally, not the least important advantage of digital files is that they can be transferred electronically within seconds, or at worst minutes for large file sizes. Compared to the effort and time required to wrap, package, and mail even a handful of invertebrates, this is an advantage that to computer-savvy people is a matter of course, but incredible and unfathomable for (usually) older scientist. To maximize the advantage, if digital files are lost, a new email can be sent right away, while a lost parcel with valuable fossils is usually irretrievable. The latter is the reason why so few institutions are willing to loan specimens. Digital files thus allow studying material far from its present physical locations, saving large sums of money otherwise spent on travel expenses for other uses, while safeguarding irreplaceable specimens.

Exhibition Use

The sole aspect of digital fossils usually noticed by the public is the use of digital files for duplication of specimens. Many fossils can be molded easily, and casts made from the moulds. Complex shapes or fragility make others nearly impossible to cast. For example, the neck vertebrae of *Giraffatitan brancai* (Janensch 1914) (formerly *Brachiosaurus brancai* Janensch 1914, see Taylor 2009) could not be mounted with the rest of the bones because their delicate laminae would have

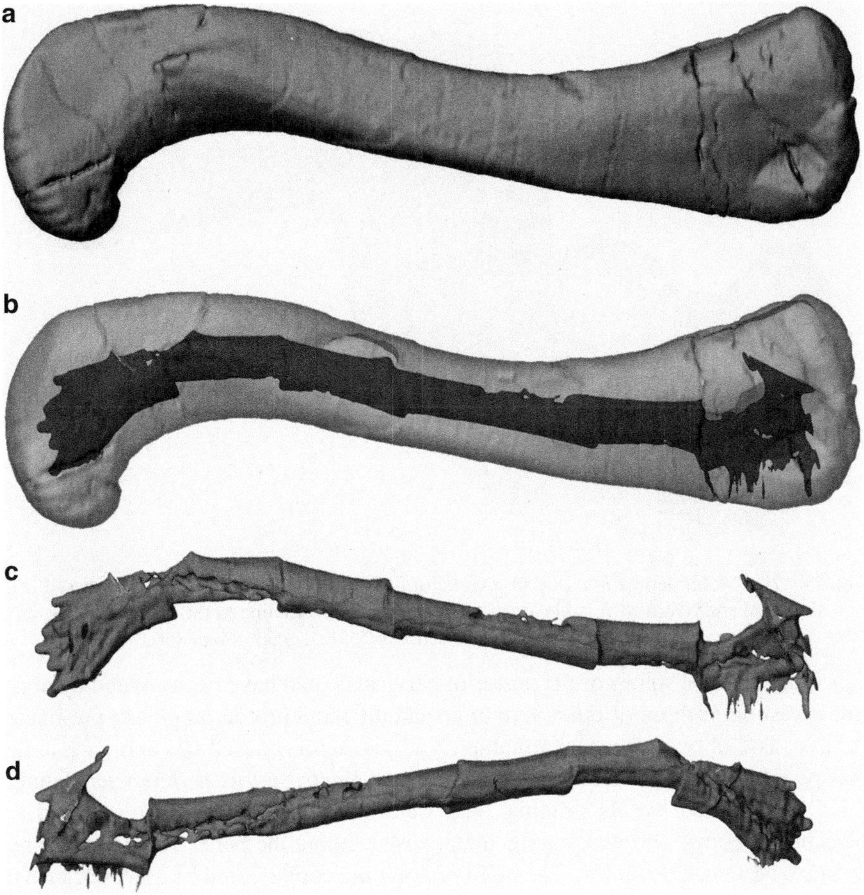

Fig. 2.3 The left femur of *Plateosaurus engelhardti* (part of GPIT/RE/7288), digital CT-scan based file. (**a**) Anterior view of the polygon mesh surface. (**b**) As (**a**) but with a transparent outer surface, showing internal surfaces relating to the marrow cavity and cracks. (**c**) The internal surfaces extracted. (**d**) As (**c**) but in posterior view. Visible in (**c**) and (**d**) is a spiraling double helix structure on the internal surface caused by two intertwined metal wires that were embedded during preparation to strengthen the bone

required so many supporting armature parts that "would distract the eye of the observer so much that the details of the highly complex external architecture would no longer be very visible" (Janensch 1950: S. 97). Casting was also impossible, because of the sheer size of the individual vertebrae, as well as the fragility of the laminae. Instead, plaster models were hand-crafted for the mount in the Museum für Naturkunde Berlin, and for a re-mounting in 2007 these were molded and casted. Here, 3D printing or other rapid prototyping techniques such as CNC milling would have allowed the rapid, albeit expensive, creation of duplicates of the original bones. Detailed laser scans could have been taken of the vertebrae, and the resulting files sectioned into parts small enough for conventional CNC milling or 3D printing.

Fig. 2.4 The left femur of *Plateosaurus engelhardti* (part of GPIT 1), digital CT-scan based file, with the distal end removed. Visible are the internal structures relating to the marrow cavity, and their connection to the outer surface where less dense rock filled cracks in the fossil bone (*arrows*)

The eroded neural spines of the posterior cervicals could have been created by non-linear scaling of those of other vertebrae, and the rapid prototyping parts pre-made for easy assembly on the steel armature. Various factors caused this approach to be dropped, but a complete copy of the stegosaur *Kentrosaurus aethiopicus* Hennig 1915 was created by 3D printing. The welding of the new armature tended to blacken the bones, and welding the metal parts without the bones caused too many errors to be a feasible option. The rapid prototyping copies could be used, because it did not matter if they were damaged. Only when the armature was completed were the real bones mounted on it.

The Naturhistorische Museum Braunschweig houses an impressive example of the application of laser scanning and rapid prototyping: the sauropod dinosaur *Spinophorosaurus nigerensis* Remes et al. 2009 was scanned and a high detail copy created for a skeletal mount. While it can be argued that the public should be shown real fossils whenever possible, this can also be achieved by displaying them as found in the field, with a skeletal mount consisting of copies above, as was the case, e.g., during the special exhibit at the Naturhistorische Museum Braunschweig that presented *Spinophorosaurus* to the public for the first time. Also, the dinosaur exhibition hall of the Museum für Naturkunde Berlin sports a cast skeleton of the theropod dinosaur *Allosaurus fragilis* (Marsh 1877), of which there is not even one real bone on exhibit. Only a femur and a tibia of a closely related, but only fragmentarily know allosaurid species is on display from the Tendaguru locality that is the focus of the exhibit. Despite this, the *Allosaurus* mount is well liked by the visitors, especially because the lightweight construction allows supporting it with only a minimum of armature visible. The future will see more displays of this kind, both because the original bones need not be exposed to the risks involved in mounting, but kept in safety, and because the original material is then much better accessible. Digital files can be used to rapid prototype bone replicas for such mounts. This is obviously also true for plaster casts or hand-made models, but both kinds of conventional replicas are often easily spotted as such, and molding large complex objects exactly has its own problems.

The flexibility of digital data is highlighted by an episode involving digital files of the wrong animal that helped re-positioning the skeletal mount of a dinosaur. *Riojasaurus incertus* Bonaparte 1967 is a prosauropod dinosaur from the Late Triassic of Argentina. When I was contacted for advice on remounting a skeleton, I was able to use files of *Plateosaurus*, scaled to the lengths given in a description of the *Riojasaurus* skeleton (Bonaparte 1971; Fig. 2.5), to test possible poses (Fig. 2.6).

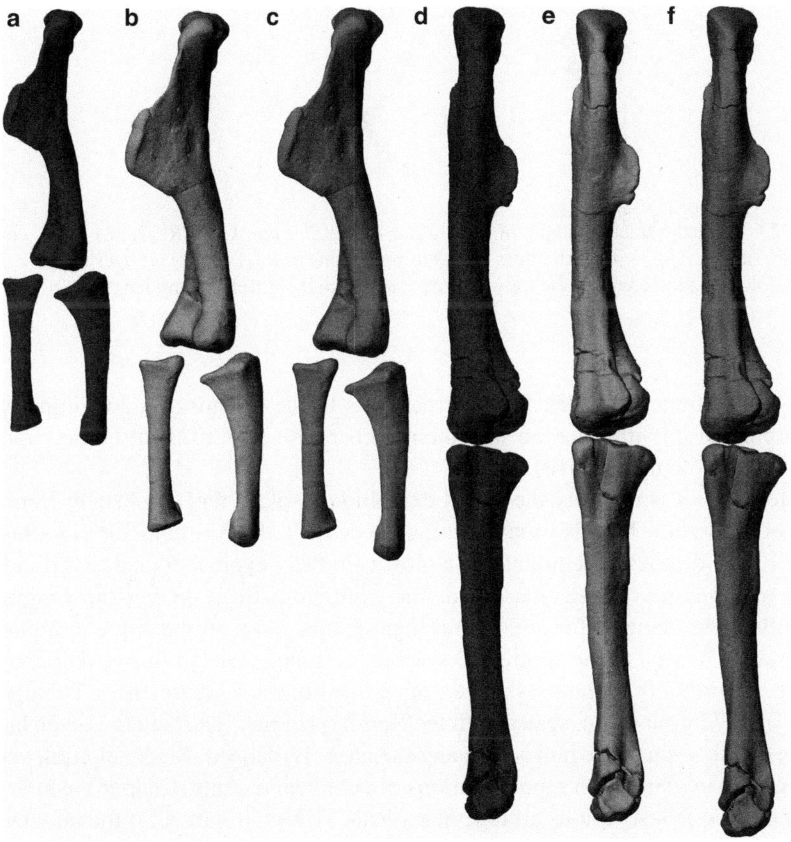

Fig. 2.5 Limb bones of *Plateosaurus engelhardti* (GPIT/RE/7288). Digital files as scanned (*dark grey*), and scaled to the proportions of *Riojasaurus incertus* for identical hindlimb length (*light grey*) and identical femur length (*medium grey*). Note the proportionally much longer humerus of *Riojasaurus*

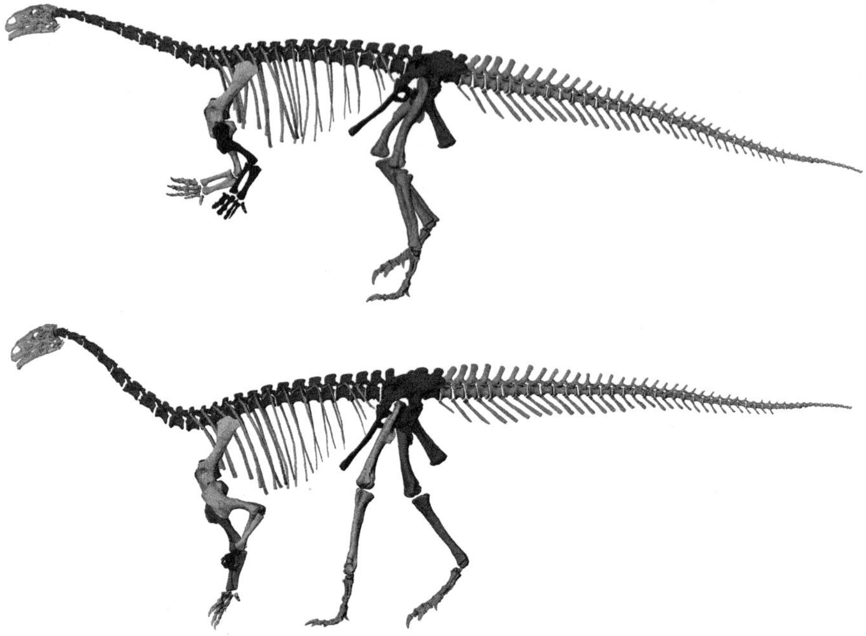

Fig. 2.6 Virtual skeletal mount of *Plateosaurus engelhardti* (GPIT/RE/7288). *Top*: correct proportions. *Bottom*: with limb bones scaled to porpotions of *Riojasaurus incertus* (see Fig. 2.5). Note how a quadrupedal posture becomes feasible due to the different limb length ratio

Of the resulting CAD files I sent screenshots to the scientists in Argentina, who incorporated this information into their decision on how to plan the new mount (Powel pers. comm. 3/2010).

However, it is not only the actual exhibition displays that can benefit from the use of digitizing, but also the planning process. If the Museum für Naturkunde Berlin ever decides to remounts its almost complete, exquisitely preserved individual of *Plateosaurus engelhardti* for the exhibition, I can deliver measurements (length, width, height) for any desirable pose, and can plan the required armature and railing as well as the spotlights down to the smallest detail in a CAD program, using my digitally mounted skeleton of the Institute for Geosciences Tübingen's GPIT/RE/7288 skeleton, scaled to fit the Berlin specimen. The benefit is even larger if the layout of an entire hall with many skeletons is planned. Relatively little effort is required to gain rough representations of complete mounts (Chap. 8), which can then be used to test various arrangements in a CAD environment. If the mounts are planned to be changed during the renovation, as was the case for the renovation of the dinosaur hall in Berlin in 2006/2007, this can also be tested using virtual skeletal mounts, provided all major bones are separately digitized. The "quick&dirty" digitizing method of choice would be mechanical digitizing after de-mounting, but laser scans can also be used. High resolution files are even sufficient to pre-plan an armature for a mount, so that parts like large support struts and main rods can be prepared to exact dimensions in advance.

Two other potential uses of digital files for exhibitions have, to my knowledge, barely been used in museums worldwide. Firstly, digital files can be used for explanatory videos or still images on computer screens, which are being set up in an increasing number of museums to supplement printed texts and audio guides. Additionally, instead of just showing fossils, or computer files of fossils, exhibits could cover evolutionary development of, e.g., ammonite lineages by morphing the digital file of an early species into that of a derived member of the family. We have come to be used to seeing such animations on TV, why not show them in museums, with the real fossils right next to them? A museum that remains static in an increasingly animated world, in which education in schools, universities and other venues uses modern techniques including animations increasingly, will sooner or later lose visitors, and fail to fulfill its mission of education.

The ability to edit digital data to create mirror images, composites or simply scaled copies is also helpful for creating displays, for example by creating scaled digital copies of vertebrae neighboring a missing one in a fossil, then 3D printing those copies to complete the mount.

Science Communication

Many researchers believe that their research speaks for itself, and will be duly noticed by colleagues and the public alike because of its quality. This is unfortunately untrue, and in recent years an increasing number of individuals and organizations have come to realize that public communication of science is an important aspect of scientific work. Here, digital files can be used interchangeably with photographs, but tend to have a higher impact, because journalists and the general public tend to be drawn to novel imagery. Also, as with museum exhibits, science published for the general public instead of for experts is not as limited by classic printing techniques, and TV shows especially offer the chance to use novel (for science) ways of presenting results and educating the public. Digital data has several advantages for presentation on the internet and on TV compared to classic photography or film of physical objects. Among them is the ability to make objects fully or partly translucent, and the ability to dissect them virtually by hiding partial surfaces, so that internal architecture can be shown easily. This can be done while the view is rotated, so that the 3D structure becomes easily understandable even on a 2D medium, the screen. Also, missing parts of a fossil can be reconstructed, and the process of reconstruction explained by switching the image between the scan of the real specimen and the reconstruction. The Museum für Naturkunde Berlin took this a step further in their redesign of the dinosaur exhibition in 2007. Here, the visitor can watch short films, in which the mounted skeletons are first filled with internal organs, next covered with musculature and then with skin, until finally the scenery changes to a reconstruction of the Jurassic landscape and the animals come alive. On the basis of these clips several paleontological techniques for soft tissue reconstruction can be explained, as well as the conclusions of biomechanical studies on the posture and locomotion of the dinosaurs. In essence, the process of making a fossil "come alive" is presented in a catchy but accurate way, which would be very hard to do without the use of digitized data.

Limitations

The biggest limitation of digital data is that by its very nature it is not the physical object it represents. It only provides the data that was collected during digitizing. Weight, density distribution, color (except when laser scans are combined with digital photographs), all internal structure (in the case of external 3D shapes), high resolution details of the surface structure, smell (although very few fossils smell), electrical properties e.g., in amber, chemical properties are usually missing in the digital file. Most of this data can be dispensed with for many research tasks, but in some cases digital representations of the fossils may not be sufficient.

Also, all digital files are processed in some way during their creation. This can induce significant errors, especially when small file sizes are used, so that results derived from the study of digital files may be useless. Thus, maximum size scans (ideally the raw data) should be used whenever possible, as e.g., Witmer and Mayle (2004) found when integrating CT and MRI data on an extant pig's knee. Every study using digital files should also include a test of file accuracy, at the very least by visual inspection and comparison of the main linear measurements of original and 3D file. If digital files are reduced in size after creation, the resulting errors should be tested for by more elaborate means. For example, the laser scan based polygon mesh files of the stegosaurian dinosaur *Kentrosaurus aethiopicus* mentioned above are too large for the entire set to be handled even by large workstation computers. Using the complete skeleton for studies on the range of motion (Mallison 2010c) would have required decreasing file size to less than 25% of the original, which leads to a process I call "digital erosion". The size-reduced files resemble bones eroded by water transport, losing sharp edges, processes and surface texture. This would have changed study results, introducing errors directly caused by the use of digital files. Thus, I was limited to using partial assemblies of the skeleton at full resolution, which increased work time and caused problems when figures of the entire skeleton had to be created.

Another limiting aspect of digitizing is its cost. Any digitizing is either time intensive, or cost intensive, or both. Depending on the method the main cost factor can be the actual scanning, because of expensive tools or high worktime demands, or the ensuing file editing. For example, high resolution laser scanning requires a scanner and scanning software, a computer (ideally a laptop) able to handle the expected file sizes, and software for follow-up editing. The number of scans that have to be taken depends on the object's size and shape, the scanner type, and the desired accuracy. Some scanners work with an object holder that rotates the object automatically, so that fewer scans are required, but this limits object size and weight. Scanning an ammonite with a diameter of 12 cm with a relatively cheap scanner took me several hours on my first try, once I had gained some experience I was able to replicate the process in about 1 h. Most of this time is not work time, because I was able to do other things while the scans were running, but the digitizing required my attention in regular intervals. The complete scanning of the *Kentrosaurus* bones one by one kept an experienced technician busy for several weeks. Afterwards, the individual scans of each bone had to be combined into one 3D surface file, a process largely automated in the scanner software. However, several files showed alignment errors, so that further editing was necessary. The time and monetary effort of scanning an entire dinosaur skeleton can only be justified if the resulting files are used repeatedly, saving time and reducing required instances of fossil handling many times.

Similarly, CT scanning is time consuming. The actual scanning is rapid, with the data transfer from the CT scanner's computer to a transfer medium being the longest part. In contrast, file extraction and editing takes much longer, and requires expensive computer programs as well as a powerful computer. Additionally, the material must be transported to the scanner, not vice versa, and fragile specimens must be adequately protected, increasing time demands and cost. If more than just the "slice" data is saved, further analyses of the raw data are possible, which again have high computational demands.

Aside from time and technical limitations, the fossils themselves sometimes make digitizing difficult or impossible. Some specimens are simply too large for some methods. Complete articulated dinosaur or mammal skeletons may not fit into even the largest CT scanners, although a handful of industrial scanners can handle specimens up to the size of a spaceship – or a hadrosaur (ScienceDaily, Dec. 3 2007; Manning et al. 2009). However, typical medical scanners are much smaller, so that even a sauropod scapula or mammoth skull may not fit through. High resolution CT scanning is even more limited, with most scanners not large enough to scan a shoe box sized fossil. Laser scanning with different scanner types can handle practically any object larger than a centimeter, but demands light reflecting surfaces. There are a number of ways to solve problems of insufficient reflexion such as coating with powder, but risk of damage may mean that they cannot be used on some fossils.

Also, there are a variety of technique-specific limitations. CT and similar techniques require differences in the density or other physical properties to exist in the specimen. If these are too small, surface extraction will fail. For example, Sutton (2008) reports that sparry calcite fossils in a largely micritic matrix could not be distinguished by HRCT and SRXTM. However, some extraordinary results have been achieved, for example calcitic fossils in limestone matrix (Dominguez et al. 2002).

Mechanical digitizing and other techniques that require physical contact with the specimens are limited to fossils stable enough to withstand touching, and care must be taken not to scratch surfaces with digitizer tips. Rough and pitted surfaces are at a greater risk, so that fragile surface features may make mechanical digitizing impossible. Also, because of the limited reach of a digitizer arm object size is only theoretically unlimited (Mallison et al. 2009), because repeated recalibrations as required for very large objects can lead to intolerable summation of errors.

Laser scanning is often combined with digital color photography to create textures for the digital files. This is supposed to add color data, and usually works quite well. However, if the specimen or scanner is moved to capture a different view, the lighting will be altered at the contact point of the two scans. This leads to either lighter or darker colors on the photograph, so that the texture usually has artificial lines where the lightness changes radically. Also, while the color difference between neighboring areas is normally well represented, except for the mentioned overlap areas of individual photographs, the color hue and saturation are usually not recorded with a proper white balance.

Laser scanning, photogrammetry and optical digitizing (and to a lesser degree mechanical and sonic digitizing with flexible digitizer arms) requires direct line of sight to the surface that is to be digitized. With a digitizer arm that has a sufficiently small tip it is sometimes possible to reach into recesses and cover obscured surfaces that cannot be seen from the outside. However, vertebrate skulls or complexly shaped vertebrae e.g., of sauropods, can require several dozens of scans and still not be fully covered.

Digitizing Methods and File Types

The digitizing methods described here are only examples for the vast number of different techniques that have been or could be applied in a paleontological context. Human ingenuity and technical progress lead almost daily to the invention of novel techniques or novel adaptations of existing techniques, so that an attempt to list all methods is doomed from the start. I have thus selected those methods that are most common in general, and most often applied in paleontology, and easily learned. Some rarely used techniques will be mentioned in passing when methods are discussed that use similar principles.

Laser Scanning

In recent years a number of affordable laser scanners have become available that can scan a field of view large enough to contain between a soda can and a packing case. Scanners that can cover a whole mounted skeleton are discussed in detail by Stoinski (Chap. 8). The simplest scanning setups for small objects are construction kit systems, using webcams, a cheap laser, a special-made background, and free computer programs (e.g., www.david-laserscanner.com). Scanners of this size range are ideal for scanning single bones of medium sized vertebrates, while those of rodents or sauropods are too small or too large, respectively. Similarly, typical large-bodies invertebrates such as mollusks are easy to scan, while single-cell organisms are usually too small. For example, my attempts to get exact scans of foraminifera of the genus *Nummulites* failed, although the largest individuals had diameters of several centimeters. The size range is roughly between 1 cm as a minimum thickness and 100 cm as the greatest length of an object, although larger objects can be scanned in pieces. Theoretically, the size range is unlimited, but accuracy suffers significantly when many scans are pieced together to create a model of a very large object.

High resolution scanners with an included high resolution digital color camera can cost several thousand to tens of thousands of dollars, and typically offer both a wider field of view and a greater measuring depth. The larger the field of view of a scanner, and the higher the resolution of a scan, the more expensive the scanner, and the longer, usually, the time for a complete scan.

Laser scanners collect data points covering the surface area of the object that is visible to the scanner. Thus, at least four scans are usually required to cover an entire object without gaps. Complex shapes, such as deep recesses, require a multitude of scans, and some surface parts may not be accessible to the scanner at all, because no direct line of sight can be established.

An important step in creating digital 3D data from laser scanning is correlating the separate scans into one file. Incorrect alignment can lead to serious errors in the final surface files. Some programs automatically recognize overlapping areas and adjust the scans accordingly, others ask the user to indicate roughly correlating points on the separate scans. More basic methods require special markings and manual adjustments. Placing of these reference markers properly requires planning and experience, because each hides part of the object from direct scanning, so that as few as possible should be used. However, this means that each marker should be visible in as many of the required separate scans as possible. Additionally, using fewer markers means that separate scans are all correlated and arranged to the same few points, so that there is no summation of errors. In contrast, when, e.g., scan 1 and scan 2 are aligned to one set of markers, then scan 3 to the two other via another set, then scan 4 by still another, the errors in each correlation will add up. This problem exists as well for other techniques that use combinations of several separate scans, such as mechanical digitizing.

There are a variety of portable and affordable laser scanner that can small object, up to the size of a soda can. Prices vary with the available resolution, scan size, scan depth, accuracy and quality of the software supplied, and a factor often overlooked: the ability to scan dark or shiny, reflecting surfaces. This point may seem irrelevant for paleontology – we typically do not scan volcanic glass or polished meteorites, nor metallic minerals or mirrors. However, shininess already begins with fossils consisting of any crystals visible to the naked eye, such as calcite and aragonite in many invertebrates. Mollusk shells that retain the original mother of pearl structure of their shell are also shiny. Usually, cheaper scanners have a less powerful laser. This means that they have a harder time getting a proper reflection of a surface that reflects laser light badly (dark materials) or away from the scanner (reflective surfaces). High power scanners, in contrast, may even allow using mirrors to scan surfaces otherwise hidden from direct view, as S. Stoinski and I found (see Chap. 8).

One solution to this problem is covering the object with a substance that reflects light well, and back at the scanner, and can later be removed easily and without damaging the object. Some companies sell powder-sticks or other tools for this purpose with their laser scanners. Alternatively, for larger objects, it may sometimes be sufficient to slightly alter the angle between scanner and object. Also, smooth and level surfaces can simply be covered with paper stickers. Scanning with strong external light, e.g., outside, may require shading of the object (e.g., Adams et al. 2010).

The smaller the object of a scan, the more important are correct registration of the separate scans, because artifacts resulting from alignment errors are proportionally larger compared to the total size of the object. The same is true for scan resolution; smaller objects require higher resolution. This means that objects with parts of different sizes, such as vertebrate skeletons, may require the use of two different scanners in order to create a complete digital model at sufficient resolution in an acceptable time frame.

High quality (and high prize) laser scanners allow scanning entire buildings, sauropod skeletons or mounted whales in one piece. Details can be found in Chap. 8. A problem that often occurs when very large paleontological objects are scanned is that the scan resolution cannot be sufficient, due to time and data size constraints, to allow simple meshing into 3D bodies of high accuracy. For example, a full skeleton scan of a medium-sized dinosaurs may be run at a resolution of 5 mm (distance between points), or even at 2.5 or 1 mm. Anything more detailed just takes too long. Neither resolution is sufficient, however, to create anything more than a rough model of the bone shapes. Thus, laser scanning of very large and complexly shaped objects provides data helpful for certain tasks, such as mass estimates based on digital 3D models derived from the scans (Gunga et al. 2007, 2008; Bates et al. 2009a, b). For more detailed work, the resolution is often insufficient, or the data too large even for high-end PCs. I experienced this problem when attempting to use a scan of *Giraffatitan brancai* (MFN mount) provided by S. Stoinski (see Chap. 8) to correct the minor errors in the mount before creating a 3D CAD model of the animal's external shape. The full scale point cloud crashed the PC, and meshing reduced versions did not give sufficient anatomical detail to adjust the vertebral column as desired. Working with the naked, un-meshed point cloud decimated to 50% of the points was not possible either, because it is impossible to estimate the depth of point clouds. This makes it nearly impossible to arrange elements in 3D. However, the solution to this problem is simple: more computing power, especially in the graphics card, would allow using the full resolution scan.

Mechanical Digitizing

Mechanical digitizing refers to the use of a robotic arm that is automatically or, more commonly, manually guided across the specimen, while the position of its tip is recorded as a 3D coordinate. Mechanical digitizing is on average cheaper than laser or CT scanning, and especially suitable for collecting landmark data, i.e., a number of specific points on an object, instead of a complete scan of its surface.

However, at limited resolutions, mechanical digitizing can also be used for surface representation, and the resulting files can be as accurate as those produced by low-resolution laser scanning, or files from high-resolution scanning that have been reduced in size (Mallison et al. 2009). Given the limitations in computing power of an average office computer, mechanical digitizing files have the same resolution as the largest laser scan files that can be used if larger assemblies such as vertebrate skeletons are investigated. My colleagues and I have detailed surface representation techniques for mechanical digitizing on the example on vertebrate bones, using a small, non-motorized digitizer arms (Mallison et al. 2009), thus only a short overview is given here.

Surface Representation

There are two main approaches to the accurate 3D representation of surfaces with a mechanical digitizer. Point clouds comparable to those created by laser scanning can be collected, from which 3D polygon meshes can be easily computed. In order to get complete representations of an object, digitizing from all sides is required, which demands multiple calibrations of the digitizer. For this, a coordinate set must be marked on the object, and extreme care must be taken that re-calibration after moving the object is accurate, otherwise the separate point clouds digitized from different sides will not match in 3D. The resulting errors are in principle the same that can occur when laser scan point clouds are not perfectly aligned. As mentioned for laser scanning, aligning should be achieved using as few separate sets of calibration marks as possible, in order to avoid summation of alignment errors.

The main advantage of point cloud digitizing with a mechanical digitizer is the ability to avoid unnecessary overlap between partial point clouds, and adjust digitizing density to match the surface topography of the object. Flat surfaces can be sampled at a far smaller resolution than highly complex structures. Also, because calibration is achieved by using coordinates, overlap between partial point clouds can be minimal. This leads to less data in the first place, and eliminates unnecessarily small mesh triangles in the overlap area, further reducing file size. For laser scans, the same reduction requires at least one extra calculation step, which reduces redundant areas of the full data set. Additionally, because digitizer arms usually feature several highly mobile joints, direct line of sight between the digitizer base and the sampled surface is not required. Instead, the operator can reach around the object, sampling complex surface topographies such as recesses or sharp keels in one go, greatly reducing the number of separate digitizing instances compared to laser scanning. This reduces the problem of inter-scan correlation, further lightening post-scanning calculation time. Overall, it is possible to achieve comparably accurate final files of 60–80% the site of a laser scan, with fewer scanning instances, and for significantly smaller file sizes during digitizing.

The second method for surface digitizing involves the recording of NURBS curves that wrap around the object. NURBS stands for non-linear uniform rational

B-splines, a type of mathematical equation that defines curves in 3D space that are mathematically easy to handle. NURBS modeling allows easy editing of curves and 3D bodies. The process of digitizing these curves is the mathematical equivalent of wrapping wires around the object to get a proxy for its shape, then pulling a cloth taught over the wires. The individual curves do not need to be exactly parallel, but can curve following surface structures, as i.e., the curves are not planar long as they do not cross each other. The curves also need not be closed, so that partial surfaces are created. These need to be combined into one complete surface later, as e.g., described by Wilhite (2002). Curves reaching fully around the object allow creating a complete representation of the fossil in one go. This requires that the digitizer arm can reach all the way around the object, and makes special supports necessary (Mallison et al. 2009). However, these can usually be produced at low cost. The operator's freedom to space and bend curves depending on the complexity of the topography allows dense sampling where required, and extremely limited sampling for areas with simple shapes such as scapula blades. The resulting files can therefore be much smaller than even low resolution laser scan files, so that large numbers of surfaces, e.g., entire skeletal mounts of large vertebrates, can be handled at the same time.

In principle, surface creation from NURBS curves follows the same methods as for CT or MRI scans, connecting corresponding points on stacked curves, but in contrast to MRI and CT data, the mechanically digitized curves are not flat (i.e., each curve point may differ from all other curve points in three, not two coordinates), and neighboring curves are not necessarily parallel.

Like all other digitizing methods, NURBS curve and point cloud digitizing both allow combining data from several specimens, even of different size or handedness. As long as sufficient overlapping area is available, or at least three easily identifiable corresponding point, data from one specimen can be scaled and mirrored to match that of others.

Landmark Analysis

Digitizing landmarks with a mechanical digitizer arm (e.g., Engels 2007) is simpler and faster than any other method and far less error prone. Collecting landmark data from photographs (e.g., Bonnan 2004, 2007) recovers only a 2D shape. With a mechanical digitizer the third dimension can be added, although using the full potential of the added information requires different analysis tools. As an added benefit, curves outlining the object can be quickly sketched in, while the 3D coordinates of each point can be directly exported from the digitizing program, usually as standardized formats, so that analysis programs can read them directly.

In addition to direct digitizing of landmark points, landmarks can also be selected and extracted from completed digital 3D surfaces. This method should be used with care, because of the inherent (although usually minimal) loss of accuracy.

CT and Other Tomographic Methods

Because of the size of CT scanners and similar machines, specimens need to be taken to the scanner, not vice versa. Usually, this means a trip to the local hospital, because the high cost and the demand for a separate room compliant with high safety standards due to the strong radiation produced make it impossible for most paleontological research institutions to acquire their own CT scanner. Working within the collections is thus usually not possible.

My own experience with CT scanning of vertebrate fossils has taught me that time spent preparing specimens for easier scanning and data extraction is usually time well spent. Small bones have little weight, and tend to rock on the CT scanner's slide during scanning. To avoid this they can be placed on a Styrofoam slice, into which small depressions fitting the fossils are cut. This not only allows error-free scanning, but is also a good method for avoiding damage during transport. Additionally, the density of Styrofoam is so low that extraction of the bone shape via a low threshold is possible. This means that even specimens of low density, such as carbonized plant fossils, can be separated from the surroundings easily, while separating them from the CT scanner's slide may be difficult. A Styrofoam slice that fits through the scanner allows scanning many small specimens in one scanning instance, saving time. In this case it is advisable to place a small identifier next to each specimen, of a sufficient density that it is well visible on the tomograms. I often used a metal wire bent into the shape of a number. It is thin enough not to create large artifacts, and allows identification of each bone based on a list I made when placing the bones on the Styrofoam tray.

When scanning is complete one should always ask the operator of the scanner to conduct a quick visualization of some of the data, in order to test if the scan worked as planned. In the unlikely case that something went wrong (e.g., the slices cannot be auto-aligned), the operator can re-export them from the raw data, or repeat the scan right away. Similarly, care should be taken that the data is properly named and identified in the scanner software, to avoid confusion. Additionally, when using a medical scanner, two copies of the data should be made, or an agreement reached with the scanner's operator that the raw data is kept for a day or two until one has had a chance to store a copy of the data elsewhere.

Digital File Handling and Editing

General Remarks

When dealing with digital files a few simple rules should always be adhered to. These are:

1. **Keep all data!** Raw scan data and all important intermediate steps of file editing should be kept, so that sources of errors can later be traced and corrected.

2. **Keep records!** All digitizing steps should be detailed in an accompanying file so that others can later reproduce all steps of data acquisition and file handling.
3. **Name properly**, and **include as much information as possible!** File names must include the catalogue number. Additionally, information on the genus, species and (if applicable) object type as well as important editing steps should be included. For example, a polygon mesh file of a *Plateosaurus engelhardti* left femur that was smoothed and reduced to 80% polygon number could be named Plateos_engelh_GPIT/RE/7288_sinFemur_SM_80perc.stl. "Speaking" names are especially important for files combined from several specimens, which should then include all catalogue numbers. It is not sufficient to include such data within the file, as annotations, or in accompanying text files, because experience tells that other researchers will not be able to use the exact same file formats, and most in-file notes are not included when exporting to other formats. Accompanying files are nice and helpful, but may get lost or be ignored.
4. **Never edit original files!** It may sound stupid, but if one opens a file and edits it, chances are high that one will forget to use "save as" and instead save over the original file, altering it. Make copy first, then edit the copy.

There are two main types of input data for surface creation discussed here, point clouds and NURBS curves. NURBS surfaces are created in one easy step in the CAD program, and need not be discussed further. Point clouds fall into two main categories.

1. Point clouds from laser scanning / mechanical, optical or sonic digitizing: these always depict the exterior of an object and come as partial point clouds that need to be registered to create one file.
2. Point clouds from CT scanning: these are already properly registered, and can, depending on the data extraction settings, portray external and/or internal structures. Usually, one never handles these files a point clouds, but uses the CT extraction software to directly create a surface, most commonly a polymesh file.

Surfaces from Laser Scanning/Mechanical Digitizing (Point Clouds)

Laser scanning typically delivers point clouds, and mechanical digitizing, as well as a number of other methods, can also be used.

The easiest way of turning a set of scans into a surface file is meshing as a polygon mesh (Fig. 2.7). Most scanning programs include this option, but it is often not without problems. Firstly, many scan sets are not correlated exactly enough to avoid small surface errors. Additionally, if the correlation is not fully automatic and uses markings and user input, scans are often correlated with overlapping areas aligned parallel, but not intertwined. If the distance between the two point clouds is a

b

c

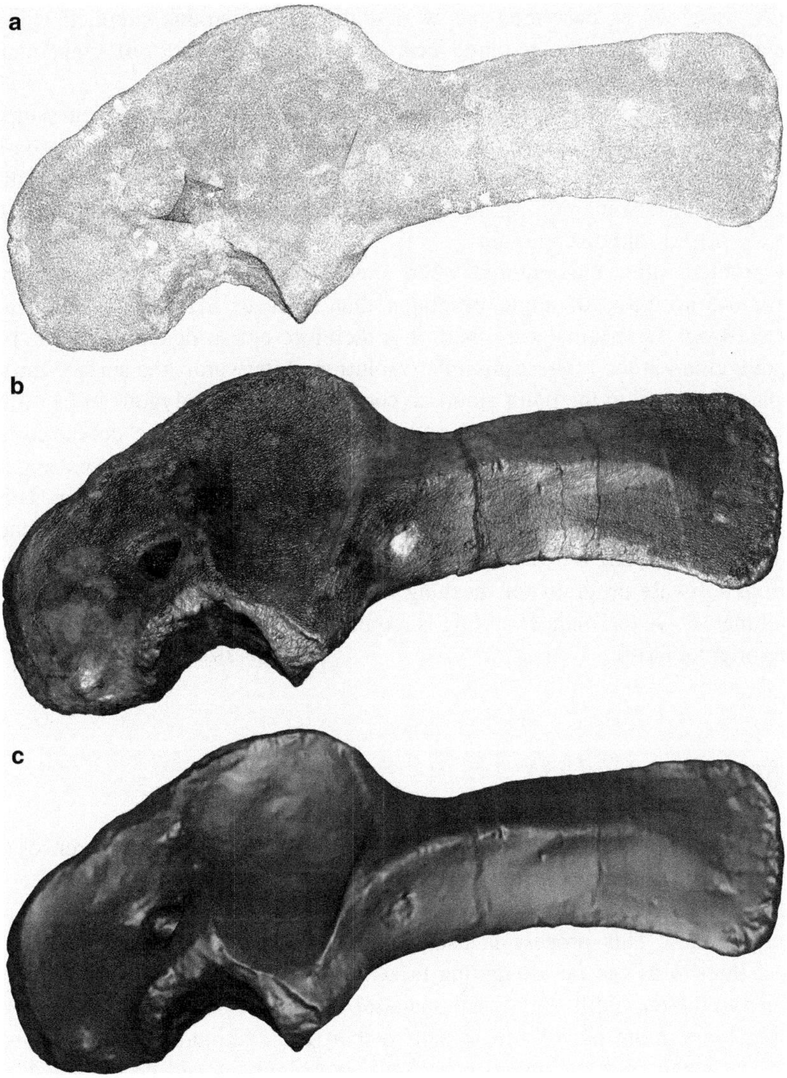

Fig. 2.7 Left scapula of *Kentrosaurus aethiopicus* (MB.R.4802) (**a**) Point cloud from high resolution laser scanning (shown in greatly reduced density) (**b**) High resolution polygon mesh with color textures (**c**) As (**b**), without color textures

much larger than the mesh triangle size, two parallel surfaces will be created. If the distance is sufficiently close to the distance between points on each scan, triangles will be created between points in both clouds. The resulting surface then looks as if it had been subjected to a shotgun blast, and requires extensive cleanup. This editing usually involves removing surface artifacts from inaccurate meshing and filling holes remaining in places where no points were scanned or the point cloud density was low, as described below. Usually, the programs supplied with the scanners can perform the required tasks, but they often lack sufficient manual controls.

Sometimes, errors in surface creation are caused by highly uneven sampling densities. This problem can be circumvented by splitting the point cloud into parts, meshing each part separately, and then combining the surfaces. The resulting surface will need some editing, though. Similarly, complex shapes may be easier to create if split into discrete units.

A problem often encountered when attempting to mesh point clouds is the insufficient accuracy of some programs that leads to 3D shapes significantly different from the original specimen. It is therefore not sufficient to simple mesh the point cloud at the highest possible resolution. Afterwards, the surface must be compared in detail to the point cloud, to check if the mesh polygons in fact follow the extreme points, e.g., along crests and on sharp spikes, or if they "cut the corner", an error likely to happen. Studies analyzing morphometric differences (e.g., the relief of teeth) depend on accurate portrayal of such extreme point (e.g., Ulhaas et al. 2004; Hansen 2006), and their quality may suffer significantly when inaccurate meshes are used. When such errors are detected it is often sufficient to use a different software program for meshing to achieve sufficiently accurate surfaces. Sometimes, however, manual editing is required to reproduce thin, sharp structures of the original fossil.

Surfaces from CT Data

Surface extraction from CT slices is based on determining density differences (i.e., grayscale value differences), and connecting those of one slice to those of the next slice that lie closest (for a more detailed description see Sutton [2008] and references therein). This process is usually not problematic. However, setting the correct thresholds can be a daunting task. In many fossils the density difference between sediment infill and fossil material is minuscule, and sometimes both densities vary more across the specimen than the maximum difference in one spot. This means that combined fossil/sediment blocks would be extracted, with parts of the fossil missing. The sole possible solution is extracting partial surfaces and combining them later. It is often worthwhile to try two thresholds, not one, so that a medium density section can be discarded.

One issue that should be addresses early on in order to save editing time later on are unwanted internal structures. Often, the mineral content of a cavity is sufficiently different from that of external sediment that it can be split from the fossil. Here, using two thresholds is also advisable. Additionally, many CT handling programs offer options to create artificial breaks, or ignoring internal, disconnected surfaces. Problematic remain density isocurves that connect the interior and exterior of a fossil, e.g., breaks in a longbone that extend to the marrow cavity. In such a case extraction of the external surface from CT data will usually lead to extraction of the internal surface, the marrow cavity wall, and the connecting duct as well (Fig. 2.4). Here, it helps if an extremely high threshold can be used, that basically separates any solid from air, so that crack infills are treated like the bone. On the other hand, the ability to digitally remove, along with the sediment, plaster infills and other repair materials can be a helpful tool for curators and preparators, too.

Resizing

Except for mechanical digitizing, all methods described here usually produce file sizes that cannot be easily handled simultaneously in large numbers by typical office and laboratory computers. A single ammonite of 10 cm diameter can result in a file of nearly 50 MB for low-resolution laser scanning, and up to 125 MB for CT or high resolution laser scanning. Even mechanical digitizing, using point clouds (the ridges on an ammonite are hard to reproduce using around-wrapping splines, so curve based digitizing is out) results in a 25 MB file. However, most research tasks can be carried out at much smaller resolution, allowing the use of much smaller files. It is, however, usually a bad idea to scan at a reduced resolution. Firstly, while it is somewhat faster, much of the scanning time is taken up by preparing and placing the specimen, adjusting the scanner position, saving files, etc. Additionally, specimens may need to be wrapped and transported – all tasks that need to be done regardless of the scanning resolution. Therefore, reducing the resolution to 50% will not result in cutting scanning time in half, but rather in a modest 10–20% time saving only. Additionally, one can never be sure what resolution and scan accuracy later research tasks may require. Who could have predicted, only 10 years ago, that 600 dpi 3D printing would become a standard application by 2010? Therefore, scanning at the maximum practical resolution (not necessarily the maximum technically possible, due to time and computing power limits) is always the best option.

If scanning at the maximum resolution is usually best, but research can usually only be carried out effectively using much smaller files, what can be done? The easiest way is to produce copies of the large files from scanning and reduce them in size until a compromise between resolution and file size is achieved that is appropriate for the task at hand. If the range of motion of a vertebrate limb is assessed (e.g., Mallison 2010b, c, in press), the fine details of surface structure of the bones are not helpful, and the digital files thus need not resolve them. For comparisons of proportional differences between different taxa, e.g. limb bone scaling within "Prosauropoda", even less detail is required, and potentially 10% file size of the original scan may be sufficient. However, such down-sized files must fulfill one important criterion: they must be accurate enough not to produce false study results. Downsizing a file is seemingly simple if it is a polymesh. Just select the proper option from the toolbar or menu, enter the target size or percentage by which to reduce, and press enter. However, this process is not always simple and reliable. Many programs offer options to preserved edges, i.e., the extremes of the 3D shapes remain supposedly untouched when the file is resized. Despite this, what happens to a shape is a process I termed "digital erosion" when encountering it on the example of *Kentrosaurus* bone files. Edges become rounded, smoothed, and appear worn down, resulting in sometimes massive alterations of the finer structures. In the case of *Kentrosaurus* these changes were unacceptable, because the files were used for a motion range analysis (Mallison 2010c), and the largest damage was evident on the zygapophyses of the vertebrae. "Digital erosion" thus altered the outcome of my analysis significantly. Care must thus be taken not to reduce files too much, and the original file and the reduced version must be compared thoroughly to ensure that the smaller file portrays the originally scanned surface accurately enough for the task at hand.

If nothing else helps, manual editing of the mesh may allow a significant size reduction, albeit for a large investment of time. The mesh must be split, with those areas that may not suffer changes at all forming separate meshes that will not be edited, while the remaining parts can be reduced in polygon number. Then, the partial meshes must be re-welded, and the contacting edges between altered and unaltered surface parts smoothed. Often, the cost in time and money of this procedure is greater than that of buying a new, more powerful computer that can handle larger file sizes. Alternatively, it may be possible to avoid resizing altogether and work with partial assemblies, as I did in the case of *Kentrosaurus*, where I studied the neck, trunk, tail and limbs in separate files.

Point cloud files can be resized by reducing the point number. In some cases it may be possible to do so without losing edge definition, but usually a surface created from a reduced point cloud shows "digital erosion" as well. NURBS curve digitizing allows simplifying the surface during its creation from the curves. If this is done in moderation the resulting error is often negligible. However, extreme caution is advisable. Even if only minor changes are made to the surface creation rules, the algorithms are prone to cause massive changes in surface curvature and thus object volume that are hardly visible to cursory inspection on the screen.

A possible way out of the dilemma of file size and file quality being inversely proportional are hybrid techniques. Mechanical (and e.g., sonic) digitizing already allows adjusting the sampling density to conform to the require resolution, reducing file size compared to a scan at full resolution. Furthermore, it may be possible to produce both high and low resolution files of an object, then cut both into parts and combine those high resolution parts that are needed with low resolution parts of areas that are not of interest. However, how such a process can be done, and whether the resulting reduction in file size is worth the effort, depends on the specific object and intended use, so that a detailed discussion here is moot.

Repairing, Splitting and Combining

An important step of laser, CT and point cloud based mechanical digitizing is the repairing and cleaning of the original scan data and the initially created surfaces.

Some programs offer simple routines that perform the tasks well on standard files. Complicated problems, such as "back-to-back" surfaces, required detailed controls and can consume large amounts of time. "Back-to-back" surfaces often occur when a thin, plate-like structure is laser scanned. Usually, the two main sides are not covered by one scan, and small errors in calibration can lead to the two surfaces intersecting, and thus coming outside-to-outside (back-to-back) with each other. This also creates a long gap along the edge of the structure. This gap cannot be closed automatically, and manual closing leads to internal surfaces extending to the outside of the file, rendering it useless for many purposes, e.g., volume determination. In such a case, the two surfaces must be deformed so that their insides become sub-parallel and face each other, and then the gap along the edge can be closed. Usually, it is easier to adjust the raw data and generate a new surface.

A repairing option that must be used cautiously is smoothing or polishing the surface. This usually is intended to remove the little pyramid shapes protruding from the surface, both outwards and inwards, caused by points lying just outside or inside the surface. Automatic options remove these reliably, but can lead to problems when such pyramids are located close to edges. Similarly, filling holes is simple, unless the holes are in places with abrupt topography changes. Here, it may be necessary to manually remove pyramids, and close the resulting and other holes by creating bridges across them. The remaining smaller holes can then be filled automatically.

Thin, plate-like structures often show a stronger version of the above mentioned pyramid errors, when two errors on the two sides connect, so that a tunnel is formed. If the surface thickness is close to the sampling distance on each surface, this error can occur so often that the surface has numerous small holes (Fig. 2.8). Such errors cannot be corrected automatically, and it may be faster to extract a new surface using slightly different parameters from the original data, or even re-scan the specimen.

Splitting and combining surfaces is usually easy as far as the technical aspects are concerned. More difficult is the question of how to scan, scale and align the parts properly, and how to document this sufficiently to make it immediately clear to any reader of a publication that the surface is not an exact representation of a fossil specimen. Usually, the absolute minimum data required for combining specimens are three clearly defined and easy to detect landmarks present on both parts that will be merged. The spatial relation between these landmarks must be unaltered, so that deformed specimens cannot be used. For the actual combining operation it is advisable to use several views at the same time, ideally six axial Cartesian views (top, bottom, left, right, front, back) plus one freely rotatable perspective view. Sadly, computer monitors of sufficient size cost more than a small car, so normally one must be content with switching between one view (for editing) and four views (one rotatable, three axial, which must be flipped between their two directions) for checking the results. The same is true for creating digital arrangements in which the surfaces stay separate, such as digital skeletal mounts.

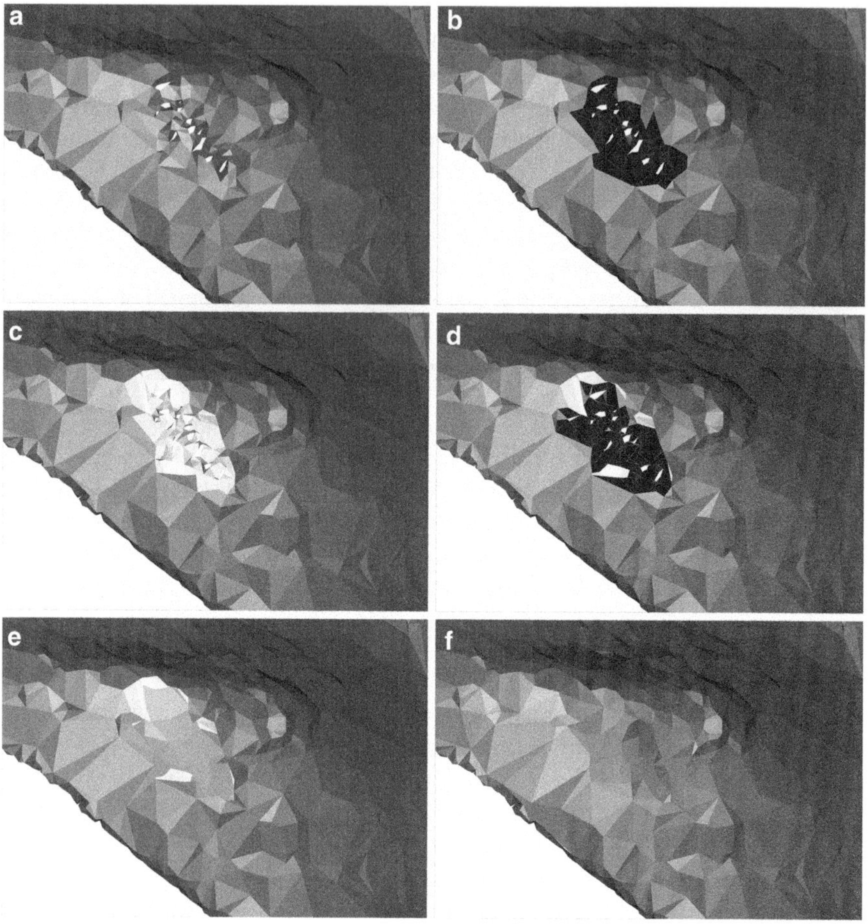

Fig. 2.8 Example for meshing error repair. CT scan based polymesh file of a *Tyrannosaurus* maxilla. The file was extracted from the slices at a very low resolution, so that the meshing distance is barely larger than the distance between the outside and inside of the bone in some places. (**a**) This has led to a number of small holes in close proximity. (**b**) To repair the triangles contacting the holes must be selected and (**c**) deleted, and (**e**), the process repeated for the other side. (**f**) Repaired surface. Note that this resembles the actual fossil more than the initially extracted surface, despite additional editing

Accuracy

How accurate a digital file represents the physical object determines what purposes it can be used for. Usually, accuracy is directly linked to scanning resolution, but later data handling can influence it negatively. For example, a high resolution laser scan point cloud may be meshed with badly selected settings, so that the size of the individual polygons is too large to create an accurate representation of the finer surface details.

File accuracy may also be influenced by elements external to the fossil that cause artifacts, e.g., by armatures or metal braces. Laser scanners have problems with reflecting or transparent materials, while metal parts (or a high content of metallic minerals) cause trouble for CT scanners. Mechanical digitizing also can face problems when fossils cannot be taken off armatures. For all methods, fixing the problems is usually easy, but requires additional time.

A significant problem for all digitizing techniques that are not touch and motion free (e.g., CT scanning, but especially mechanical digitizing) is motion by the fossil during data collection. This can lead to artifacts easily recognized, but small motions may go undetected, and induce significant errors. This is the main reason why any digital file should be checked again the original before use, e.g., by measurements of several length parameters.

The highest scan resolutions can be achieved with HR CT scanning and laser scanners. Up to approximately 10,000 points per centimeter [or ~ 24,000 dpi (dots per inch)] are possible for laser scanning, and less than 5 μ resolution for HRCT.

One especially dangerous source of error that can drastically reduce accuracy is human error during file editing. If the units of a file as set wrong (e.g., "mm" instead of "m") the mistake will be noticed and corrected. However, if a file is supposed, e.g., to be reduced to 90% of polygon number, but is rescaled for geometric size instead, the resulting error (equivalent to a erroneously scaled scale bar in a figure) may escape attention, and the results of studies based on this file may be taken at face value. Care must thus be taken that all file versions are kept, that the initial file version, the final version used for studies and the original fossil are compared, and that notes are kept of what was done to what file. Ideally, this information is included in the file name, e.g., Dactylioceras_IFGT332_50perc.stl indicating that the file of an ammonite scan was reduced to 50% polygon number.

Summary

Overall, digital models of fossils can be a helpful tool for preservation, research and education. Digital files make possible some research projects impossible on real fossils, because they allow non-destructive investigations of internal structures or (re-)deformations, greatly reduce the work load for others such as range of motion analyses for vertebrates, ease the tasks of planning museum exhibits, especially when large multi-part mounts are desired, and in some cases may be the only option for preserving fragile specimens for theoretically unlimited time.

However, digitizing is not a magical solution to all problems, and due to the potentially high cost and work effort proper planning is required to use it effectively. Many techniques are still in their infancy, and a certain degree of trial and error testing should always be allowed, because even minor changes to a technique may provide large benefits.

Researchers and curators alike should therefore publish not just the results of their research using digital files, but also the details of their digitizing technique and their experience with handling and using the files. Only if the know how is properly exchanged can full use be made of the enormous potential digitizing offers for paleontology. Files should be made available for researchers worldwide, and file transfer should be provided without hassle, e.g. through FTP servers. This requires funding, and it is up to the professional paleontological community to convince funding agencies to provide the required support.

References

Adams, T.L., Strganac, C., Polcyn, M.J. & Jacobs, L.L. 2010: High resolution three-dimensional laser-scanning of the type specimen of *Eubrontes (?) glenrosensis* Shuler, 1935, from the Comanchean (Lower Cretaceous) of Texas: Implications for digital archiving and preservation. Palaeontologia Electronica 13, 13.3.1T.

Ager, D.V. 1965: Serial grinding techniques. In: B. Kummel & D. Raup (Eds.): *Handbook of palaeontological techniques* 212-224. San Francisco: H. Freeman and Co.

Alonso, P.D., Milner, A.C., Ketcham, R.A., Cookson, M.J. & Rowe, T.B. 2004: The avian nature of the brain and inner ear of *Archaeopteryx*. Nature 430, 666–669.

Angielczyk, K.A. & Sheets, H.D. 2007: Investigation of simulated tectonic deformation in fossils using geometric morphometrics. Paleobiology 33, 125–148. doi: 10.1666/06007.1.

Balanoff, A.M., Bever, G.S. & Ikejiri, T. 2010: The braincase of *Apatosaurus* (Dinosauria: Sauropoda) based on Computed Tomogprahy of a new specimen with commens on variation and evolution in sauropod neuroanatomy. American Museum Novitates 3677, 1–32.

Bamforth, E.L. 2008: *Multibranched rangeomorphs form the Ediacaran Mistaken Point assemblage, Newfoundland, Canada*. Thesis (Master, Geological Sciences & Geological Engineering), Kingston, Canada: Queen's University.

Bamforth, E.L., Narbonne, G.M. & Anderson, M.M. 2008: Growth and ecology of an Ediacaran multibranched rangeomorph from the Mistaken Point Assemblage, Newfoundland. Journal of Paleontology 82, 763–777.

Bates, Breithaupt, B.H., Falkingham, P.L., Matthews, N.A., Hodgetts, D. & Manning, P.L. 2009: Integrated LiDAR & photogrammetric documentation of the Red Gulch Dinosaur Tracksite (Wyoming, USA). In: S.E. Foss, J.L. Cavin, T. Brown, J.I. Kirkland, & V.L. Santucci (Eds.): *Proceedings of the Eighth Conference on Fossil Resources: Utah* 101–103.

Bates, K.T., Manning, P.L., Hodgetts, D. & Sellers, W. 2009: Estimating the mass properties of dinosaurs using laser imaging and 3D computer modeling. PLoS ONe 4(2):e4532. doi: 10.1371/journal.pone.0004532.

Bednarz, M. & McIlroy, D. 2009: Three-dimensional reconstruction of "phycosiphoniform" burrows: implcations for identification of trace fossils in core. Palaeontologia Electronica 12, 12.3.13A.

Bonaparte, J.F. 1967: Dos nuevas "faunas" de reptiles triásicos de Argentina. Ameghiniana 10, 89–102.

Bonaparte, J.F. 1971: Los tetrapodos del sector superior de la Formation Los Colorados, La Rioja, Argentina (Triasico Superior). Opera lilloana 22, 1–183.

Bonnan, M.F. 2004: Morphometric analysis of humerus and femur shape in Morrison sauropods: implications for functional morphology and paleobiology. Paleobiology 30, 444–470.

Bonnan, M.F. 2007: Linear and geometric morphometric analysis of long bone sclaing patterns in Jurassic neosauropod dinosaurs: their function and paleobiological implications. Anatomical Record 290, 1089–1111.

Bookstein, F.L. 1991: *Morphometric tools for landmark data*. Cambridge: Cambridge University Press.

Boyd, A.A. & Motani, R. 2008: Three-dimensional re-evaluation of the deformation remoival technique based on "jigsaw puzzling". Palaeontologia Electronica 11.2.A.

Breuckmann, B. 1993: Bildverarbeitung und optische Meßtechnik in der industriellen Praxis. Mücnhen: Franzis Verlag.

Brief, J., Behle, J.H., Stellzig-Eisenhauer, A. & Hassfeld, S. 2006: Precision of landmark positioning on digitized models from patients with cleft lip and palate. The Cleft Palate-Craniofacial Journal 43, 168–173.

Brochu, C.A. 2000: A digitally rendered endocast for *Tyrannosaurus rex*. Journal of Vertebrate Paleontology 10, 1–6.

Brochu, C.A. 2003: Osteology of *Tyrannosaurus rex*: insights from a nearly complete skeleton and high-resolution computed tomographic analysis of the skull. Memoir of the Society of Vertebrate Paleontology 7, Journal of Vertebrate Paleontology 1–140.

Brombin, C., Mo, G., Zotti, A., Giurisato, M., Salmaso, L. & Cozzi, B. 2009: A landmark analysis-based approach to age and sex classification of the skull of the mediterranean Monk seal (*Monachus monachus*) (Hermann, 1779). Anatomia, Histologia, Embryologia 38, 382–386.

Bushman, K. & Cornell, S.R. 2009: Morphologic landmark analysis and glabellular fenestrae count of North American *Cryptolithus* trilobites. Geological Society of America, Northeast Section Meeting (Portland, ME), Abstracts with Programs Paper No. 39–10.

Carlson, W.D., Rowe, T.B., Ketcham, R.A. & Colbert, M.W. 2003: Applications of high-resolution X-ray computer tomography in petrology, meteoritics and palaeontology. In: F. Mees, R. Swennen, M. Van Geet, & P. Jacobs (Eds.): *Applications of X-ray Computer Tomography in the Geosciences*, Special Publications 215, 7–22. London: Geological Society.

Christian, A. & Dzemski, G. 2007: Reconstruction of the cervical skeleton posture of *Brachiosaurus brancai* Janensch 1914 by an analysis of the intervertebral stress along the neck and a comparison with the results of different approaches. Fossil Record 10, 38–49.

Clark, N.D.L., Adams, C., Lawton, T., Cruickshank, A.R.I. & Wood, K. 2004: The Elgin marvel: using magnetic resonance imaging to look at a mouldic fossil from the Permian of Elgin, Scotland, UK. Magnetic Resonance Imaging 22, 269–273. doi: 10.1016/j.mri.2003.09.006.

Clark, S. & Morrison, I. 1994: CT scan of fossils. In: P. Leiggi & P. May (Eds.): *Vertebrate paleontological techniques*. Cambridge: Cambridge University Press.

DeQuardo, J.R. 1999: Landmark analysis of corpus callosum shape in schizophrenia. Biological Psychiatry 46, 1712.

DeVore, M.L., Kenrick, P., Pigg, K.B. & Ketcham, R.A. 2006: Utility of high resolution x-ray computed tomography (HRXCT) for paleobotanical studies: an example using London Clay fruits and seeds. American Journal of Botany 93, 1848–1851.

Dierick, M., Cnudde, V., Masschaele, B., Vlassenbroeck, J., Von Hoorebeke, L. & Jacobs, P. 2007: Micro-CT of fossils preserved in amber. Nuclear Instruments and Methods in Physics Research Section A: Accelerators, Spectrometers, Detectors and Associated Equipment. Proceedings of the 10th International Symposium on Radiation Physics - ISRP 10 580, 641–643.

Dominguez, P., Jacobson, A.G. & Jefferies, R.P.S. 2002: Paired gill slits in a fossil with a calcite skeleton. Nature 417, 841–844. doi: 10.1038/nature00805.

Donoghue, P.C., Bengtson, S., Dong, X.P., Gostling, N.J., Huldtgren, T., Cunningham, J.A., Yin, C., Yue, Z., Peng, F. & Stampanoni, M. 2006: Synchrotron X-ray tomographic microscopy of fossil embryos. Nature 442, 680–683.

Engels, S. 2007: *Funktionelle Morphologie des Schädels und der Bezahung der Ursidae*. Diploma Thesis, Frankfurt: Johann Wolfgang Goethe Universität Frankfurt am Main. Abgerufen von http://hopsea.mnhn.fr/pc/thesis/Diploma_Engels_2007.pdf.

Elewa, A.M.T. 2007: A powerful electronic journal in the new millennium. Palaeontologia Electronica Vol. 10, Issue 1; 2A:2p, 240KB; http://palaeo-electronica.org/paleo/2007_1/editoria/focus.htm

Elewa, A.M.T. 2009: The role of paleontology in the "Electronic Era". Journal of Geology and Mining Research Vol. 1(10)

Elewa, A.M.T. 2010: Paleoinformatics: The superhighway to modern paleontology. Journal of Geology and Mining Research Vol. 2(1)

Falk, D., Hildebolt, C., Smith, K., Morwood, M.J., Sutikna, T., Jatmiko, Saptomo, E.W., Imhof, H., Seidler, H. & Prior, F. 2007: Brain shape in human microcepalics and *Homo floresiensis*.

Proceedings of the National Academy of Sciences USA 104, 2513–2518. doi: 10.1073/pnas.0609185104.

Fink, W.L. & Humphries, J.H. 2010: Morphological description of the extinct North American sucker *Moxostoma lacerum* (Ostariophysi, Catostomidae), based on High-Resolution X-ray Computer Tomography. Copeia 2010, 5–13. doi: 10.1643/CI-09-089.

Fraser, N.C., Olsen, P.E., Dooley, A.C.J. & Ryan, T. 2007: A new gliding tetrapod (Diapsida: ? Archosauromorpha) from the Upper Triassic (Carnian) of Virginia. Journal of Vertebrate Paleontology 12, 261–265. doi: 10.1671/0272-4634(2007)27[261:ANGTDA]2.0.CO;2.

Gehling, J., Narbonne, G.M. & Anderson, M.M. 2000: The first named Ediacaran body fossil, *Aspidella terranovica*. Palaeontology 43, 427–456.

Gower, J.C. & Dijksterhuis, G.B. 2004: *Procrustes problems*. Oxford: Oxford University Press.

Gunga, H., Suthau, T., Bellmann, A., Friedrich, A., Schwanebeck, T., Stoinski, S., Trippel, T., Kirsch, K. & Hellwich, O. 2007: Body mass estimations for *Plateosaurus engelhardti* using laser scanning and 3D reconstruction methods. Naturwissenschaften 94, 623–630.

Gunga, H., Suthau, T., Bellmann, A., Stoinski, S., Friedrich, A., Trippel, T., Kirsch, K. & Hellwich, O. 2008: A new body mass estimation of *Brachiosaurus brancai* Janensch 1914 mounted and exhibited at the Museum of Natural History (Berlin, Germany). Fossil Record 11, 28-36.

Hansen, J. 2006: *Untersuchung zur funktionellen Struktur der carnassialen und postcarnasialen Zahnreihe im Viverridengebiss*. Diploma Thesis, Frankfurt: Johann Wolfgang Goethe Universität Frankfurt am Main. Abgerufen von http://hopsea.mnhn.fr/pc/thesis/Diploma_Hansen_2006.pdf.

Haubitz, B., Prokop, M., Doehring, W., Ostrom, J.H. & Wellnhofer, P. 1988: Computed tomography of *Archaeopteryx*. Paleobiology 14, 206–213.

Hauboldt, H. 1990: Ein neuer Dinosaurier (Ornithischia, Thyreophora) aus dem Unteren Jura des nördlichen Mitteleuropa. Revue de Paléobiologie 9, 149–177.

Hounsfield, G.N. 1980: Computed Medical Imaging. Science 210, 22–28.

Hughes, N.C. & Jell, P.A. 1992: A statistical/computer graphic technique for assessing variation in tectonically deformed fossils and its application to Cambrian trilobites from Kashmir. Lethaia 25, 317–330.

Hutchinson, J.R., Anderson, F.C., Blemker, S. & Delp, S.L. 2005: Analysis of hindlimb muscle moment arms in *Tyrannosaurus rex* using a three-dimensional musculoskeletal computer model: implications for stance, gait, and speed. Paleobiology 31, 676–701.

Hutchinson, J.R. & Garcia, M. 2002: *Tyrannosaurus* was not a fast runner. Nature 451, 1018–1021.

Hutchinson, J.R., Miller, C.E., Fritsch, G. & Hildebrandt, T. 2008: The anatomical foundation for multidisciplinary studies of animal limb function: examples from dinosaur and elephant limb imaging studies. In: H. Endo & R. Frey (Eds.): *Anatomical Imaging Techniques: Towards a New Morphology* 23–38. Berlin: Springer.

Ichaso, A.R., Dalrymple, R.W. & Narbonne, G.M. 2007: Paleoenvironmental and basin analysis of the late Neoproterozoic (Ediacaran) upper Conception and St. John's groups, wet Conception Bay, Newfoundland. Canadian Journal of Earth Sicences 44, 25–41.

Janensch, W. 1914: Übersicht über der Wirbeltierfauna der Tendaguru-Schichten nebst einer kurzen Charakterisierung der neu aufgeführten Arten von Sauropoden. Archiv für Biontologie 3, 81–110.

Janensch, W. 1950: Die Skelettrekonstruktion von *Brachiosaurus brancai*. Palaeontographica Supplement 7, 97–103.

Kak, A.C. & Slaney, M. 2001: *Principles of computerized tomographic imaging*. Philadelphia: Society of Industrial and Applied Mathematics.

Kermack, D.M. 1970: True serial thin-sectioning of fossil material. Biological Journal of the Linnean Society 2, 47–53. doi: 10.1111/j.1095-8312.1970.tb01686.x.

Kielan-Jaworowska, Z., Presley, R. & Poplin, C. 1986: The cranial vascular system in taeniolabidoid multituberculate mammals. Philosophical Transactions of the Royal Society of London 313, 525–602. doi: 10.1098/rstb.1986.0055.

Krey, K. & Dannhauer, K. 2008: Morphometric analysis of facial profile in adults. Journal of Orofacial Orthopedics / Fortschritte der Kieferorthopädie 69, 424–436.

Kullmer, O., Huck, M., Engel, K., Schrenk, F. & Bromage, T. 2002: Hominid tooth pattern database (HOTPAD) derived from optical 3D topometry. In: B. Mafart & H. Delingette (Eds.): *Three-Dimensional Imaging in Paleoanthropology and Prehistoric Archaeology. Acts of the XIVth UISPP Congress, University of Liège, Belgium, 2-8 September 2001*. British Archaeological Record International Series 1049 71–82.

Lohmann, G.P. 1983: Eigenshape analysis of microfossils: A general morphometric procedure for describing changes in shape. Mathematical Geology 15, 659–672.

Luo, Z. & Ketten, D.R. 1991: CT scanning and computerized reconstruction of the inner ear of multituberculate mammals. Journal of Vertebrate Paleontology 11, 220–228.

MacLeod, N.M. and Patterson, R.T. 1998: The Role and the Promise of Electronic Publishing in Paleontology. Palaeontologia Electronica 1(1).

Mallison, H. 2007: *Virtual dinosaurs - developing Computer Aided Design and Computer Aided Engineering modeling methods for vertebrate paleontology*. Doctoral Thesis, Eberhard-Karls-Universität Tübingen. Accesses September 4, 2009, at http://tobias-lib.ub.uni-tuebingen.de/volltexte/2007/2868/.

Mallison, H. in press: *Plateosaurus* in 3D: How CAD models and kinetic/dynamic modeling help bringing an extinct animal to life. In: N. Klein, K. Remes, C. Gee, & P.M. Sander (Eds.): *Biology of the Sauropod Dinosaurs: Understanding the life of giants*, Life of the Past Series. Bloomington: Indiana University Press.

Mallison, H. 2010a: The digital *Plateosaurus* I: body mass, mass distribution, and posture assessed using CAD and CAE on a digitally mounted complete skeleton. Palaeontologia Electronica 13, 13.2.8A. Accessed January 25, 2011, at http://www.palaeo-electronica.org/2010_2/198/index.html.

Mallison, H. 2010b: The digital *Plateosaurus* II: an assessment of the range of motion of the limbs and vertebral column and of previous reconstructions using a digital skeletal mount. Acta Palaeontologica Polonica 55, 433–458. doi: 10.4202/app.2009.0075.

Mallison, H. 2010c: CAD assessment of the posture and range of motion of *Kentrosaurus aethiopicus* Hennig 1915. Swiss Journal of Geosciences 203, 211–233. doi: 10.1007/s00015-010-0024-2.

Mallison, H., Hohloch, A. & Pfretzschner, H. 2009: Mechanical digitizing for paleontology - new and improved techniques. Palaeontologia Electronica 12. Accessed September 4, 2009, at http://palaeo-electronica.org/2009_2/185/index.html.

Manning, P.L., Morris, P.M., McMahon, A., Jones, E., Gize, A., Macquaker, J.H.S., Wolff, G., Thompson, A., Marshall, J., Taylor, K.G., Lyson, T., Gaskell, S., Reamtong, O., Sellers, W.I., van Dongen, B.E., Buckley, M. & Wogelius, R.A. 2009: Mineralized soft-tissue structure and chemistry in a mummified hadrosaur from the Hell Creek Formation, North Dakota (USA). Proceedings of the Royal Society of London, Series B: Biological Sciences 276, 3429–3437.

Mardia, K.V. & Dryden, I.L. 1998: *Statistical shape analysis*. Chichester: Wiley.

Marsh, O.C. 1877: Notice of new dinosaurian reptiles from the Jurassic Formation. American Journal of Science, 3rd Series 14, 514–516.

McLean, D., Robinson, J. & Archer, M. 2001: Artefact reduction on CT images of fossils to allow 3D visualisation. Radiation Physics and Chemistry 61, 723–724.

Meyer, H.V. 1837: Mitteilung, an Prof. Bronn gerichtet [*Plateosaurus engelhardti*]. Neues Jahrbuch für Mineralogie, Geologie und Paläontologie 1837, 817.

Meyer, H.V. 1861: *Archaeopteryx litographica* (Vogel-Feder) und *Pterodactylus* von Solenhofen. Neues Jahrbuch für Mineralogie, Geognosie, Geologie und Petrefakten-Kunde 1861, 678–679.

Muir-Wood, H.M. 1934: On the internal structure of some Mesozoic Brachiopoda. Philosophical Transactions of the Royal Society of London 505, 511–567. doi: 10.1098/rstb.1934.0012.

Newman, A. 1998: Pyrite oxidation and museum collections: a review of theory and conservation treatments. Geological Curator 6, 363–371.

Osborn, H.F. 1905: *Tyrannosaurus* and other Cretaceous carnivorous dinosaurs. Bulletin of the American Museum of Natural History New York 21, 259–265.

Osborn, H.F. 1924: Three new Theropoda, Protoceratops zone, central Mongolia. American Museum Novitates 144, 1–12.

Raichlen, D.A., Gordon, A.D., Harcourt-Smith, W.E.H., Foster, D.A. & Haas, W.R.J. 2010: Laetoli footprints preserve earliest direct evidence of human-like bipedal biomechanics. PLoS ONE 5, e9769. doi: 10.1371/journal.pone.0009769.

Remes, K., Ortega, F., Fierro, I., Joger, U., Kosma, R. & Ferrer, J.M.M. 2009: A new basal sauropod dinosaur from the Middle Jurassic of Niger and the early evolution of sauropoda. PLoS ONE 4, e6924. doi: 10.1371/journal.pone.0006924.

Rixon, A.E. 1976: The effects of the decomposition of iron pyrites within a specimen and methods used for its arrest. In: A.E. Rixon (Ed.): *Fossil Animal Remains: Their Preparation and Conservation* 139-152. London: Athlone Press.

Rowe, T.B., Colbert, M., Ketcham, R.A., Maisano, J. & Owen, P. 2001: High-resolution X-ray computed tomography in vertebrate morphology. Journal of Morphology 248, 277–278.

Rushton, A.W. & Smith, W. 1993: Retrodeformation of fossils - a simple technique. Palaeontology 36, 927–930.

Sampson, S.D. & Witmer, L.M. 2007: Craniofacial anatomy of *Majungasaurus crenatissimus* (Theropoda: Abelisauridae) from the Late Cretaceous of Madagascar. Memoir of the Society of Vertebrate Paleontology 8, Journal of Vertebrate Paleontology 27 (Supplement to No. 2), 32–102.

Schwarz, D., Vontobel, P.L., Eberhard, H., Meyer, C.A. & Bongartz, G. 2005: Neutron tomography of internal structures of vertebrate remains: a comparison with X-ray computed tomography. Palaeontologia Electronica 8, http://palaeo-electronica.org/2005_2/neutron/issue2_05.htm.

ScienceDaily. 2007: December 3: Dinosaur Mummy Found With Fossilized Skin And Soft Tissues. http://www.sciencedaily.com/releases/2007/12/071203103349.htm.

Seilacher, A. 1999: Biomat-related lifestyles in the Precambiran. Palaios 14, 86–93.

Sellers, W.I., Manning, P.L., Lyson, T., Stevens, K.A. & Margetts, L. 2009: Virtual palaeontology: Gait reconstruction of extinct vertebrates using hgh performance computing. Palaeontologia Electronica 12, 12.3.13A.

Shah, J. & Srivastava, D.C. 2007: Strain estimation from distorted vertebrate fossils: application of the Wellman method. Geological Magazine 144, 211-216. doi: 10.1017/S0016756806002962.

Simpson, G.G. 1933: A simplified serial sectioning technique for the study of fossils. American Museum Novitates 634, 1–6.

Smith, S.Y., Collinson, M.E., Rudall, P.J., Simpson, D.A., Marone, F. & Stampanoni, M. 2009: Virtual taphonomy using synchrotron tomographic microscopy reveals cryptic features and internal structure of modern and fossil plants. Proceedings of the National Academy of Sciences USA 106, 12013–12018. doi: 10.1073/pnas.0901468106.

Sollas, I.B. & Sollas, W.J. 1913: A study of the skull of a *Dicynodon* by means of serial sections. Philosophical Transactions of the Royal Society of London 204, 201–225. doi: 10.1098/rstb.1914.0006.

Sollas, W.J. 1904: A method for the investigation of fossils by serial section. Philosophical Transactions of the Royal Society of London 196, 259–265. doi: 10.1098/rstb.1904.0008.

Srivastava, D.C. & Shah, J. 2006: Digital method for strain estimation and retrodeformation of bilaterally symmetric fossils. Geology 34, 593–596. doi: 10.1130/G22374.1.

Stevens, K.A. & Parrish, J.M. 1999: Neck posture and feeding habits of two Jurassic sauropod dinosaurs. Science 284, 798–800.

Stevens, K.A. & Parrish, J.M. 2005a: Digital reconstructions of sauropod dinosaurs and implications for feeding. In: K.A. Curry-Rogers & J.A. Wilson (Eds.): *The sauropods: Evolution and Paleobiology* 178-200. Berkeley: University of California Press.

Stevens, K.A. & Parrish, J.M. 2005b: Neck posture, dentition, and feeding strategies in Jurassic sauropod dinosaurs. In: K. Carpenter & V. Tidwell (Eds.): *Thunder Lizards: The Sauropodomorph Dinosaurs* 212–232. Bloomington: Indiana University Press.

Stooshnov, A. & Buttler, C.J. 2001: The treatment of specimen labels affected by pyrite decay. Geological Curator 7, 175–180.

Subsol, G., Mafart, B., Silvestre, A. & De Lumley, M. 2002: 3D image processing for the study of the evolution of the shape of the human skull. Presentation of the tools and preliminary results. In: B. Mafart & H. Delingette (Eds.): *Three-Dimensional Imaging in Paleoanthropology and Prehistoric Archaeology. Acts of the XIVth UISPP Congress, University of Liège, Belgium, 2–8 September 2001*. British Archaeological Record International Series 1049 37–45.

Sutton, M.D. 2008: Tomographic techniques for the study of exceptionally preserved fossils. Proceedings of the Royal Society B: Biological Sciences 275, 1587–1593. doi: 10.1098/rspb.2008.0263.

Sutton, M.D., Briggs, D.E.G., Siveter, David J. & Siveter, Derek J. 2001: Methodologies for the visualization and reconstruction of three-dimensional fossils from the Siluarian Herefordshire Lagerstätte. Palaeontologia Electronica 4, http://palaeo-electronica.org/2001_1/s2/issue1_01.htm.

Taylor, M.P. 2009: A re-evaluation of *Brachiosaurus altithorax* Riggs 1903 (Dinosauria, Sauropoda) and its generic separation from *Giraffatitan brancai* (Janensh 1914). Journal of Vertebrate Paleontology, 29, 787–806.

Ulhaas, L., Kullmer, O., Schrenk, F. & Henke, W. 2004: A new 3-d approach to determine functional morphology of cercopithecoid molars. Annals of Anatomy 186, 487–493.

Wedel, M.J. 2003a: The evolution of vertebral pneumaticity in sauropod dinosaurs. Journal of Vertebrate Paleontology 23, 344–357.

Wedel, M.J. 2003b: Vertebral pneumaticity, ais sacs, and the physiology of sauropod dinosaurs. Paleobiology 29, 243–255.

Wilhite, D.R. 2002: Digitizing large fossil skeletal elements for three-dimensional applications. Palaeontologia Electronica 5, http://palaeo-electronica.org/2002_2/scan/issue2_02.htm.

Witmer, L.M. & Mayle, H. 2004: The best of both worlds. Integrating CT and MR. Image 17, 17–19.

Witmer, L.M. & Ridgely, R.C. 2008: The paranasal air sinuses of predatory and armored dinosaurs (Archosauria: Theropoda and Ankylosauria) and their contribution to cephalic architecture. Anatomical Record 291, 1362–1388.

Wood, D.A., Dalrymple, R.W., Narbonne, G.M., Gehling, J. & Clapham, M.E. 2003: Palaeoenvironmental analysis of the late Neoproterozoic Mistaken Point and Trepassey formations, Southeastern Newfoundland. Canadian Journal of Earth Sicences 40, 1375-1391. doi: 10.1139/e03-048.

Zollikofer, C.P.E., de Léon, M.S.P., Lieberman, D.E., Guy, F., Pilbeam, D., Likius, A., Mackaye, H.T., Vigneaud, P. & Brunet, M. 2005: Virtual cranial reconstruction of *Sahelanthropus tchadensis*. Nature 434, 755–759.

Zollikofer, C.P.E., de Léon, M.S.P., Martin, R.D. & Stucki, P. 1995: Neanderthal computer skulls. Nature 375, 283–285.

Chapter 3
Paleoinformatics: Past, Present and Future Perspectives

Jane K. Dolven and Hans Skjerpen

Introduction

The idea for making the World-Wide Web was created by Tim Berners-Lee already in the late 1980s, and further developed together with Robert Cailliau in 1990. It was designed as a system of interlinked hypertext where the philosophy was to have an "open, nonproprietary and free of charge" access where everyone could contribute with information regardless of location. The importance of this global system of interconnected computer networks (Internet) grew throughout the 1990s along with an incredible amount of information added by people everywhere. As the introduction of computers in the workspace gradually moved information from being analog to being digital, the evolution of the World-Wide Web gradually moved information from being local to being global. The Web became the mass medium of the twenty-first century.

Discovering the fast growing Internet soon also initiated the idea of having databases online. The benefits proved to be many. Presenting data online, whether general or more specific, gave the potential of reaching a new and more geographically dispersed audience. Besides being, as regular databases, important for storing information and preventing it from getting lost, online databases (i.e., databases with a web interface) provide excellent ways of sharing information between people especially when the web interface allows querying for data.

Such systems were, as in many other fields and disciplines, soon adapted in paleontology, and the term *paleoinformatics* was coined. Paleoinformatics includes the use of information technology to manage, preserve and distribute paleontological data. The term was first introduced during the international Senckenberg conference in Frankfurt 1997 dealing with "Fossils and the future; paleontology in the 21st

J.K. Dolven (✉)
Department of Geosciences, University of Oslo, PO Box 1047, Blindern, Oslo 0316, Norway
e-mail: jane@radiolaria.org

H. Skjerpen
Virtual One AS, 0491 Oslo, Norway
e-mail: hans@one.no

A.M.T. Elewa (ed.), *Computational Paleontology*,
DOI 10.1007/978-3-642-16271-8_3, © Springer-Verlag Berlin Heidelberg 2011

century" (MacLeod and Guralnick 2000; MacLeod et al. 2000). The overall consensus was that the use of paleoinformatics could help the paleontological community to better structure and access data as well as preserve the systematic information and expertise that is fundamental to the field. With the introduction of computers in the workspace in the 1980s, paleontological institutions (universities, museums, surveys) as well as individual researchers gradually started digitizing their paleontological data and experimenting with paleontological species databases (e.g., Riedel 1989). Along with the evolution of the World-Wide Web during the 1990s came the possibility to put this information online, first as simple static text-pages that was later accompanied or replaced by "primitive" online databases, leading up to the present where the majority of web pages are some form of advanced online database where the pages are generated from structured data and presented according to the user's requests.

The following text will reflect on different aspects of paleoinformatics based on the authors' experience with an online database for the radiolarian community. We admit our experience is limited and make no claim to be paleoinformatics experts, but we share our story and perspectives in the hope that they may be found useful. We close with some thoughts on the future of paleoinformatics.

Paleontological Internet Databases

Although it is already a decade since paleoinformatics was introduced to the field of paleontology there is still a large amount of paleontological data (including published literature and collections) that needs to be digitized, structured and made available online.

Many private, public and academic initiatives are making great progress on digitizing publications (e.g., Biodiversity Heritage Library and Google Scholar), but there is still much work to be done. Digitizing publications helps saving old and important legacy work for the generations to come. But the different ways of digitizing information provides different levels of usefulness. Scanning a page from an old book preserves an image for the future, but does not allow text-searches. OCR'ing the scanned image adds another level of usefulness, transforming the image into searchable text. Extracting the information in the text into a standard structure in a database provides yet another level of utility, allowing the data to be used in concert with other data. However, the ultimate goal is reached when putting the digitized information online with an open access.

Before starting to build online systems one need to choose the most appropriate structure for the data. Many information systems start out as local solutions to specific problems for a single researcher or a group of people. A few of these local tools evolve into larger systems for departments, organizations or faculties. And in turn a few of these "make it" on to the Internet. This technical version of "the survival of the fittest" is a natural and healthy mechanism, and contributes to keeping different tools and systems at their appropriate level. What systems "qualifying for" inclusion in the cyberinfrastructure will of course differ from field to field, but in general the more basic and fundamental the data is the better it is suited for infrastructure inclusion. Science invariably builds its pyramid of knowledge from the basic data and upwards towards the current hypothesis and findings. Paleoinformatics is the use of information technology on these pyramids, and cyberinfrastructure can be viewed as the collection of interconnected systems each containing parts of the pyramid of knowledge. There may be several pyramids of knowledge in a field (sometimes reflecting conflicting views). Integrating data from different pyramids may cause some problems due to different ways of structuring and categorization information in the different systems. The requirement for exchanging data in meaningful ways between systems is the standardization of the exchange protocols, which in practice boils down to having compatible data structures. Preparing for the future by conforming to standards if existing, even for local systems, will be an investment and of a large value to the field. If no standards exist, the development of standards for the most basic data should be a priority superseded only by the work of digitizing legacy data.

The primary reason for putting a database online is to provide access to data. Although read-only access to data is the most common, there are more and more systems experimenting with giving write access to registered users. The web 2.0 movement of user generated content online is a model that may be well suited to Academia, by letting potentially all researchers from a field share and divide the work of building the community knowledge by adding what they know into existing online databases, at any time, and from anywhere. Wikipedia is maybe the most well known example of this model. Examples of more specialized web based database systems (portals) include Tree of life (www. tolweb.org) and Encyclopedia of Life (www.eol.org).

Challenges in Paleoinformatics

Most paleontologists do not have advanced programming skills or necessary knowledge of database modeling. This competency must normally be acquired elsewhere and often at a cost which requires funding. Unfortunately there is often little or no incentive for paleontologists to spend their time or budgets building online databases and contributing to the cyberinfrastructure. This is, we believe, mainly a result of how most scientists are evaluated, i.e., mainly based on number of publications and citations. Many researchers are due to an already tight schedule afraid of losing valuable "publishing time" if participating in online database initiatives. Others are concerned about putting unpublished data online and risk "losing a future publication". Perhaps if scientists would be evaluated not only by the number of publications, but on other accomplishments as well, like contributions to their fields building and maintaining valuable online databases, we think there would be more time spent on paleoinformatics.

One of the main challenges for paleoinformatics is therefore to get paleontologists as well as funding agencies interested, enthusiastic and committed. A possible solution may be to get the researchers involved in making systems that make their own work easier, i.e., systems that help organize their own information in a way that make them more effective, and as a useful side-effect makes the work of other people easier as well. Those are the systems that have the most chance for use, contribution and success. In sosiodynamic terms they may be described as systems with "a critical mass of one user".

Beyond the challenges of getting people to participate and contribute lies at least two other major issues that needs to be addressed: data quality and copyright issues. Before starting building a database one have to choose which level of quality will be required for the data from the future users. In other words whether the database will just be a collections of all available information on a subject regardless of whether it is formally correct or not, or whether the database needs to consist of high quality data. One way of improving the latter is to have experts in the field to add the information, dedicated web-editors that controls the input data, and/or have the data being based on peer-reviewed published publications. This leads to the second problem namely copyright issues. Many publishing companies have extremely strict copyright policies. They are dependent on selling their publications to survive and are therefore not interested in their exclusive information being available somewhere else for free, even though it ironically in many cases are the scientist themselves that have produced the data, written the paper/book and sometimes even paid for getting it published. Many open access journals, where the authors retain all rights to his/her research, may change this.

Radiolaria.org: A Case Study

The authors' first venture into the field of Paleoinformatics was quite coincidental. On a yearlong research exchange program in 1999–2000, the need arose for some way of sharing information and in particular comparing radiolarian species from different geographical locations between fellow scientists. The Web was fairly new, and few other online catalogues of radiolarians existed. Having the combined skill set between us, both Paleontology and Informatics, the idea was born to create an online database ourselves that would be useful in our work. The idea was simple: Create a web-based catalogue of radiolarians containing species names, images, descriptions, synonyms, references along with their geographical/geological distribution. This would then be available from campus, from home, and from anywhere with a computer connected to the Internet. In addition to being accessible from anywhere, we decided to have a web interface that enabled logging in to the system and adding, modifying and deleting content. It was a very simple system and the prototype was developed in a few weeks as "proof of concept". The design process was iterative with constant evaluations of how the system could help with the task at hand, and although we saw the potential broader use, the tool started out as being of local use.

We called the prototype Radiolaria.org (www.radiolaria.org) and after showing it to a few colleagues we were encouraged to present the system at the upcoming international radiolarian conference (InterRad IX) in Blairsden, California, in September 2000. The system was presented as a possible collaborative tool for the radiolarian community, and although a few where somewhat skeptical to the expected life span of this unfunded brainchild of some junior researcher, the participants at InterRad provided valuable feedback on what was needed and what would be useful. Radiolaria.org has since then grown into a useful resource for radiolarian researchers and been expanded to include much more than only taxonomic information (Dolven and Skjerpen 2006). Several radiolarian specialists have been contributing filling the database with valuable information and giving feedback on how to improve the system. It now also holds, in addition to taxonomic information, a large reference database with several thousand references as well as forums for posting and answering questions. It has become the official site for the International Radiolarian Association (InterRad) with information regarding memberships, officers, newsletters and conferences. Our work with Radiolaria.org made us realize the importance of working together as a group and providing a meeting place for the community built by and for the radiolarian people.

Some Perspectives and Approaches

Different groups have different needs, and may therefore require different solutions to their unique mix of problems. These needs may also change as the system evolves. Below follows a few of our perspectives that might also be of value to others.

Radiolaria.org was built to create a simple catalog to link species names, images, description and synonyms/references. It was made for the radiolarian community to fill their needs. The advantage of building small systems is that they *do one thing well* and require less resources, both people and money wise, to get up and running and be maintained than larger systems.

It may seem tempting to fill the database with whatever information you have at hand regardless of quality and think that you will check the content and correct mistakes later. Quantity before quality may of course work for some systems, but in our experience a system is better served if *all data is quality assured* before entered or shortly afterwards by web editors. We also believe the database is better off when a minimum set of mandatory properties is added, in the sense that there are something entered for every property. In our case that means that a species is not published online without having at least the species name, one description (preferably the original description), one image, one synonym/reference and a tagged geographical/geological distribution.

Any website should aim to be easy to learn and easy to use. We believe a few activities are essential in creating user-friendly systems, i.e., listening to your users and doing simple usability testing having users trying to perform tasks on the website while observing them (without giving directions or hints). This simple form of user testing will give valuable insights into how people think. Letting colleagues from your department or acquaintances on workshops/conferences test the system and solve defined tasks is a great way of finding out what parts work and which do not. Make it easy for your web users to send you feedback via e-mail or an online feedback-form. This is an easy way of identifying problem areas so the system can be improved. Read the comments and suggestions, and use them to make your own decisions.

Being open is an important part of the spirit of the Internet. From open (free of charge) access to information to open source code, the Internet has embraced the ideas of sharing and performing community work. As parts of a larger cyberinfrastructure there are two types of openness that are important apart from the obvious openness to access: open to integration and open to contributions. By enabling other servers to run queries on your data and receive structured results, you open up for integration. By letting other researchers (or servers) add to or modify content in the database, you open up to contributions. Although opening up comes with the cost of higher demands on security and quality control, it is a requirement for furthering the field of paleoinformatics.

Paleoinformatics for the Future

The problems of building a cyberinfrastructure, where data can flow freely between different information systems, are shared between many fields of science. The emerging solutions are thus also of common use, and the field of paleontology can benefit from applying them as they become the standard way of building information infrastructures.

One of the basic requirements for all cyberinfrastructure includes standardizing the structure of data. In research you often want to compare your data with other researcher's data. Data cannot be compared unless they have the same structure and follow conventions. In practice for the field of paleontology this means developing and adopting defined structures for species and related information (e.g., geographical/geological occurrence, diversity). Some initiatives, like "Darwincore" (http://rs.tdwg.org/dwc/index.htm) are working on this. Elewa (2010) suggests calling such systems "Taxonomic Information Systems".

The second requirement is to make the data available according to the standards. On the Internet (X)HTML is the standard for creating web pages, and the XML-format has become the standard way of both defining the structure of data (Document Type Definition) and exchanging structured data (XML-file). While XML is a format well suited for processing by a computer, it is not very readable by humans. Traditional web pages, while readable by humans, do not have the content structured enough to be processed by a computer. Emerging solutions for making data available to both computers and humans are being developed that enables data on a standard web-page to be structured (with tags) in a way that does not interfere with the readability by humans, but still enables a computer to extract data in its structured form. Some of the most popular markups at present are "microdata", "microformats" and "RDFa".

The third and final requirement for enabling data to flow freely, but still controlled by its authors and owners, is licensing the data for easy and legal distribution. Creative Commons licenses enable the distribution of data while controlling modification, redistribution and commercial use.

By following the requirements of making structured data available according to a standard and with proper licensing, any initiative will contribute to its field in a very scalable and robust way. The web pages containing the structured data can and will be picked up by search-engines making them findable by both humans and computers. The search engines and Internet archives companioned with the licensing of the data will make the data accessible and reduce the dependency of personal commitments or available funding and thereby grant the contribution of shared information eternal life.

Concluding Remarks

Working with Paleoinformatics and online databases is an exercise in many different skills, requiring specialists and generalists in different fields. The work involves, in addition to paleontological expertise, everything from the detective work of tracking old legacy data, the sharp eye needed for good digitalization work, the politics of standardization processes and rights management, the analytics of selecting or developing databases and computer applications, the insights and patience needed for usability testing, to the psychology and leadership needed to make everyone and everything come together.

The cross-disciplinary nature of such projects provides (at least the potential of) rich learning experiences outside their own fields for everyone involved.

Acknowledgement Many thanks to prof. Ashraf Elewa and prof. Kjell R. Bjørklund for reviewing this manuscript and providing useful comments and suggestions.

References

Dolven, J.K. and Skjerpen, H., 2006. An online micropaleontology database: Radiolaria.org. Eclogae Geologicae Helvetiae, 99: 63-66.

Elewa, A.M.T., 2010. Paleoinformatics: The Superhighway to Modern Paleontology. Journal of Geology and Mining Research, 2(1).

MacLeod, N. and Guralnick, R., 2000. Paleoinformatics. In: Lane, R.H., Steininger, F., Kaesler, R. L., Ziegler, W., Lipps, J.H. (editors), Fossils and the future; paleontology in the 21st century. Senckenberg-Buch, 74: 31-36.

MacLeod, N., Diver, P., Guralnick, R., Lazarus, D. and Malmgren, B., 2000. Computers, quantification, and databases. In: Lane, R.H., Steininger, F., Kaesler, R.L., Ziegler, W., Lipps, J.H. (editors), Fossils and the future; paleontology in the 21st century. Senckenberg-Buch, 74: 191-201.

Riedel, W.R., 1989. Identify: a Prolog program to help identify fossils. Computers & Geosciences, 15(5): 809-823.

Chapter 4
Calculating the Tempo of Morphological Evolution: Rates of Discrete Character Change in a Phylogenetic Context

Stephen L. Brusatte

Introduction

Paleontologists and biologists are often interested in the tempo of evolution: how fast or slow does evolution proceed? There are many separate components of evolution – the development and extinction of lineages, molecular change, and morphological transformation are three of the most general – and these may or may not be related to each other. Therefore, it is instructive to look at each of these separately (if possible), in order to gain a more nuanced understanding of evolutionary change. This chapter will focus specifically on morphological evolution: changes in size, shape, and discrete anatomical features and how rates of change can be calculated. Such calculations have a rich legacy in the macroevolution literature (e.g., Westoll 1949; Derstler 1982; Forey 1988; Cloutier 1991; Ruta et al. 2006; Brusatte et al. 2008a).

Understanding the patterns and tempo of morphological change may give unique insights into large-scale evolutionary processes. For instance, the theory of adaptive radiation suggests that organisms undergo a high amount and rate of morphological change as they rapidly speciate into numerous lineages after colonizing a new island or ecosystem (e.g., Schluter 2000; Gavrilets and Losos 2009). On longer time scales, the hypothesis of punctuated equilibrium holds that long periods of morphological stasis are interrupted by episodes of dramatic morphological change (Gould and Eldredge 1977; Stanley 1979). In order to test whether a certain clade radiated adaptively or underwent punctuated episodes of change, it is first necessary to quantitatively measure morphological rates of change. More broadly, it is simply interesting to establish patterns: are certain clades or time intervals characterized by more morphological change, or higher rates of change, than others? Once patterns are robustly established, they may be marshaled as evidence in favor of certain

S.L. Brusatte
Division of Paleontology American Museum of Natural History, Central Park West at 79th Street, New York, NY 10024, USA
and
Department of Earth and Environmental Sciences, Columbia University, New York, NY, USA
e-mail: sbrusatte@amnh.org

A.M.T. Elewa (ed.), *Computational Paleontology*,
DOI 10.1007/978-3-642-16271-8_4,

evolutionary processes or used to describe the large-scale narrative of clade history (e.g., Ruta et al. 2006; Brusatte et al. 2008a).

This chapter is intended as a primer on morphological rates analysis. Using a sample dataset, presented by Brusatte et al. (2008a) and focused on the evolution of dinosaurs and other archosaurian reptiles, I will provide a step-by-step guide to calculating rates of discrete morphological character change in a phylogenetic context. These calculations require three basic components: a database of morphological characters scored for a range of species-level taxa, a phylogenetic tree showing the relationships of those taxa, and information on the absolute ages of each taxon. Armed with this information, amounts of character evolution per branch of the phylogeny can easily be calculated, and these can be converted to rates by dividing by the time duration of the branch in question. Once a rate is calculated for each branch, these can be binned according to clade or time interval, and statistical tests can determine whether differences between these bins are significant. When these steps are completed, the patterns of morphological rate are then available for description and interpretation.

Required Information

Database of Morphological Features

The first step in calculating morphological rates is to compile information on the morphology of the organisms in question. The goal here is to represent something as complex as organismal morphology in a manageable dataset that encompasses the maximum amount of information possible. There are several possible approaches, and each is commonly used in studies of morphological evolution. Researchers may build datasets based on numerous measurements, such as the length and thickness of individual bones and ratios of the sizes of certain body parts to others (e.g., McGowan and Dyke 2007). Alternatively, morphometric techniques quantify the shape of whole organisms or individual components, either in two or three dimensions (e.g., Bookstein 1991; Elewa 2004; Zelditch et al. 2004). Both of these techniques use continuous data: size and shape, respectively, which vary in a continuum. A different approach is to utilize discrete characters. These are not continuous, as each organism can only be scored for one or multiple distinct states. Examples of discrete characters include the presence/absence of certain structures or the possession of a specific type of structure (say, an enlarged tooth instead of a small tooth).

Which approach to follow depends intimately on the group being studied and the questions being addressed. Many invertebrate groups are ideal for morphometric or continuous size analysis, as their morphology is quite simple and thus amenable to representation by a table of measurements or shape outlines in two or three dimensions. However, when morphology becomes more complex discrete characters are often a better approach, as they allow more flexibility in documenting features that may be masked in simple size or shape compilations. For instance, the fantastic antlers of moose or extreme cranial kinesis of snakes would be completely ignored by a table of size measurements. Even morphometric techniques, which can be exceedingly powerful in quantifying shape, are difficult to use when studying vertebrates with numerous individual bones of different shapes and sizes. Most vertebrate bones are too complex to represent in two dimensions, and the time and computational effort needed to do a three-dimensional morphometric analysis of each bone across several taxa are extraordinary. Therefore, discrete characters are an ideal representation method for vertebrate morphology, and have been used in numerous studies (e.g., Jernvall et al. 1996; Ruta et al. 2006; Brusatte et al. 2008a, b; Ruta 2009). The downside of using discrete characters is that they must be compiled by the researcher, which requires subjective decisions about which characters to include and often intensive hands-on work with specimens. The upside, however, is that cladistic datasets are usually comprised of discrete characters, resulting in a readily available pool of data.

The example analysis profiled here, the study of archosaur morphological rates presented by Brusatte et al. (2008a), is based on an extensive discrete character dataset that was built by combining information from published cladistic character sets with additional data. This dataset includes 437 characters scored across 64 Triassic taxa, three of which are non-archosaur outgroups. Most of these characters are unordered – that is, if there are more than two discrete states then it only requires a single "step" to transition from any one state to any other. Ten of these characters, however, were ordered, which requires multiple evolutionary steps (character changes) to pass between states. For instance, it would cost one step to pass from state 1 to 2 but five steps to pass from state 1 to 6. These steps are important because it is the number of character changes per branch that is ultimately counted in the rate calculation.

An example of discrete character usage is shown in Table 4.1, which depicts the first five characters in the analysis and their scores in a sample of the archosaur taxa. The full dataset can be found in the supplementary information of Brusatte et al. (2008a). Figure 4.1 provides a schematic illustration of discrete character states on actual archosaur specimens.

Phylogenetic Tree

With a set of morphological features in hand, the next task is to examine how these features changed during the evolutionary history of the group being studied. After all, we are interested in the rate of morphological *change*. More specifically, how many transitions between different character states occurred and when, and in what groups, did these changes happen? These questions require a phylogenetic context, because it is necessary to know (or at least to have a good estimate of) the genealogical relationships of the organisms being studied in order to determine those exact places on the family tree where one character state changed into another.

Table 4.1 Discrete character dataset

Taxon	1	2	3	4	5
Scleromochlus	1	0	1	0	1
Silesaurus	0	0	?	0	0
Herrerasaurus	0	0	0	0	0
Plateosaurus	0	0	0	0	1
Paleorhinus	1	0	0	0	0
Aetosaurus	0	0	0	0	1
Hesperosuchus	0	0	0	0	0
Ornithosuchus	0	0	0	0	0
Rauisuchus	?	?	0	?	?
Lotosaurus	0	0	0	1	1

An example of the type of characters that can be used for morphological rates analysis and their scores for a series of taxa. These are the first five characters from the Triassic archosaur dataset of Brusatte et al. (2008a), and the taxa are a small sample of the 64 genera in the original dataset

1. Skull, length: less than (0) or greater than (1) 50% length of presacral column
2. Antorbital fenestra, shape: elliptical or circular (0); triangular, with elongate and narrow anterior point (1)
3. Orbit, anteroposterior length: less (0) or greater (1) than 25% skull length
4. Orbit, shape: circular or elliptical (0); tall and narrow, with maximum height more than 1.5 times maximum width (1)
5. External naris, length of longest dimension: less (0) or greater (1) than longest dimension of antorbital fenestra

The phylogeny itself can be obtained using many different techniques. The most common method, at least in vertebrate paleontology research, is to conduct a parsimony analysis of discrete morphological characters, usually using a computer program such as PAUP* (Swofford 2000) or TNT (Goloboff et al. 2008). Other methods for phylogeny reconstruction include maximum likelihood and Bayesian techniques, both of which are model-based approaches that rely on an assumed model of evolution instead of simply minimizing the number of character changes as in a parsimony analysis (Huelsenbeck et al. 2001; Felsenstein 2004). The characters used to reconstruct the phylogeny may be the same characters used to quantify morphology, as discussed above. However, it is important to exclude from a phylogenetic analysis any characters that may be correlated to another character, but when quantifying the broad range of morphologies in a group for rates analysis such characters may be beneficial. Furthermore, although autapomorphies (characters unique to the terminal taxa) are usually excluded in phylogenetic analyses, they are important to include in a rates analysis since rates are also calculated for the branches leading to the terminal taxa. Therefore, researchers should carefully consider which characters to include in a phylogenetic analysis, and not blindly use a dataset constructed to represent total morphology. Guidelines for the construction and use of phylogenetic characters, including potential pitfalls, are summarized in many sources (e.g., Kitching et al. 1998; Felsenstein 2004; Schuh and Brower 2009).

Brusatte et al. (2008a) utilized a single, resolved phylogeny of Triassic archosaurs that stemmed from a higher-level phylogenetic analysis of the clade (Brusatte 2007; Brusatte et al. 2010). The phylogenetic analysis recovered a series of most

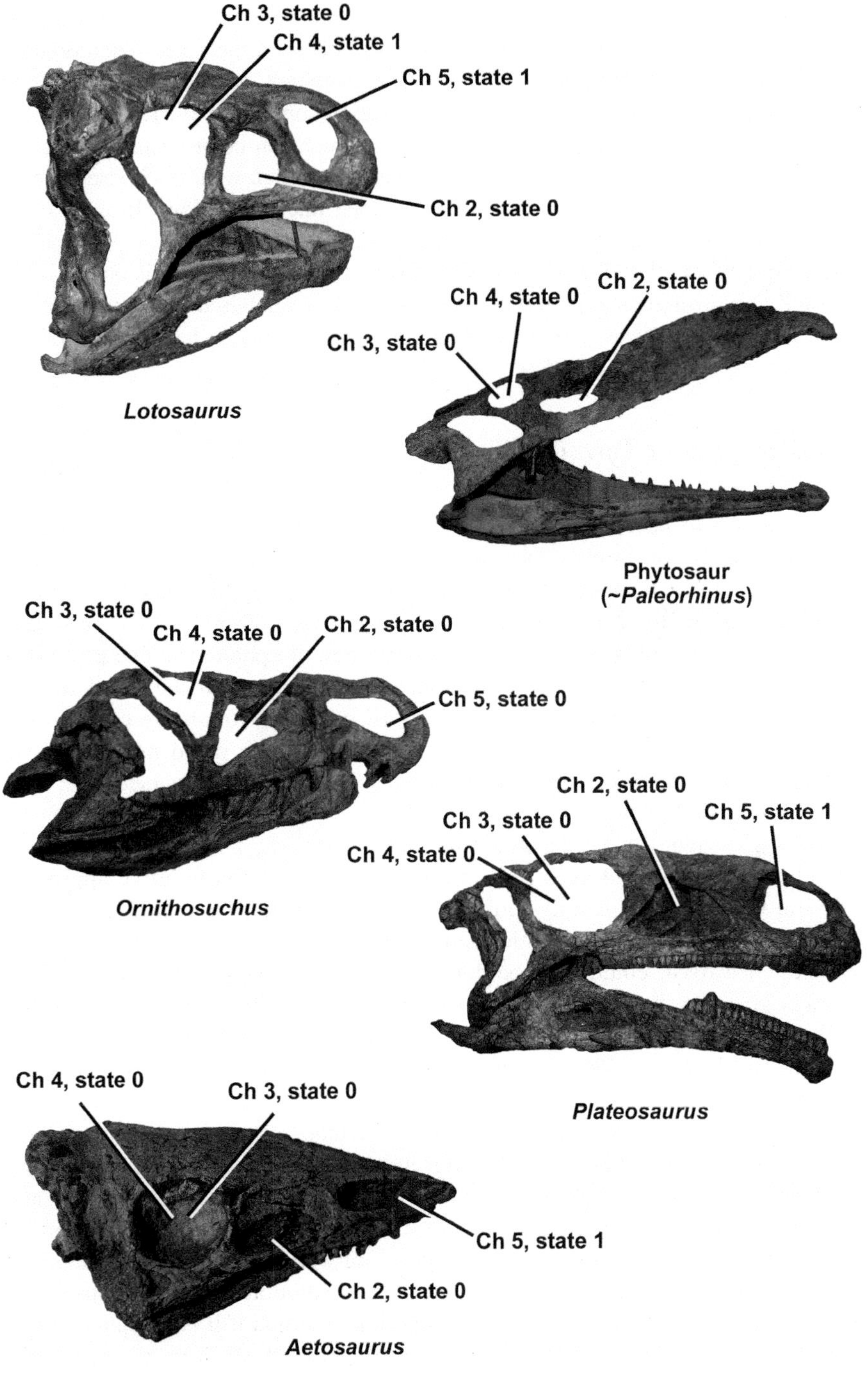

Fig. 4.1 Schematic illustration showing the presence of discrete character states on a sample of archosaurian reptile specimens. The characters being assessed are characters 2–5 in Table 4.1, and the discrete scores for each of these five taxa are listed in the data matrix in Table 4.1

parsimonious (optimal) trees: fully resolved trees with the same minimal number of character changes but with different relationships between the taxa. When doing a phylogenetic study these trees would usually be combined into a consensus tree that shows the major patterns shared between the various optimal trees. However, these consensus trees will usually include polytomies, which can be problematic for rates analysis. Because of this conundrum, researchers may often want to choose a single resolved tree, or a subset of resolved trees, for their rate calculations. Brusatte et al. (2008a) followed a straightforward approach, in which they simply used the first optimal tree reported by the phylogeny reconstruction software. This is one possible choice if only a single tree is needed, but researchers are advised to use a sample of trees in their rates analysis, in order to gauge and correct for biases due to tree topology.

Absolute Ages of Terminal Taxa

Finally, it is also necessary to have absolute ages for each taxon being studied. Rates analysis is not simply concerned with the amount of morphological change, but the speed of such change. Rates are amounts divided by time, and the denominator is just as important as the numerator in this equation. Therefore, it is necessary to quantify the ages of taxa with as much care as their morphological features. What we are interested in is the first appearance of each taxon in the fossil record. However, this may be problematic, as many (if not most) fossils are only loosely dated and are not known from strata that can be directly dated by radiometric techniques. Oftentimes the age of a fossil is only estimated to a coarse range, such as "Late Triassic."

Although vague, these designations are not useless, and can be converted into absolute, numerical ages by a straightforward and conservative approach. Brusatte et al. (2008a) simply took the finest age resolution available for each fossil and then assigned the midpoint of that range as the absolute age. For instance, if a fossil could only be dated to the Late Triassic, then the absolute age was given as the midpoint of that range (218.3 million years old according to Walker and Geissman 2009). If another fossil was more finely resolved to the Carnian, however, its age was considered to be the midpoint of that stage (231.5 million years old).

While the above approach is easy to implement other techniques are also possible, and may often be preferable. For instance, multiple ages (e.g., minimum, maximum, and intermediate of the finest age resolution) can be assigned and different rates eventually calculated, in order to better gauge biases caused by imprecise dating. More sophisticated computer software can consider a randomized draw of absolute ages pulled from the finest age resolution of the first appearance (e.g., Pol and Norell 2006), which is the optimum method if it is computationally feasible for the researcher. In light of these different methods, it is important to be explicit about techniques used and the source of geological and paleontological data supporting individual age assessments.

Morphological Rates Calculation

When the above information – morphological character dataset, phylogenetic tree, absolute ages for terminals – is compiled, then the process of calculating rates can begin. This is a multi-step process, but all calculations and statistical analyses are straightforward and only a limited set of readily-available software programs are necessary. More sophisticated programming knowledge may simplify many of these steps, but the basics can be mastered by anyone with a phylogenetic reconstruction program, a calculator, and a basic computer statistical package.

Determining the Number of Character Changes per Branch

The first major step is to calculate how many character changes happened on each branch of the phylogeny. From here on out, the various calculations will focus on each *branch* of the tree, not the observed (terminal) taxa whose morphology was represented in the character dataset. Each terminal taxon has its own branch, of course, but cladograms also include internal branches that link hypothetical common ancestors. These branches are also included in the rate calculations, because characters changed along them and they existed for an absolute duration of time.

Determining the number of character changes along each branch – the "amount" of evolution that happened on that branch – relies on character optimization. All of the morphological characters observed in the terminal taxa are mapped onto the tree, showing where these characters changed from one state to another (in other words, on which branch of tree this change took place). There are numerous strategies for optimizing characters onto a phylogeny, including parsimony, likelihood, and Bayesian techniques (e.g., Felsenstein 2004; Schuh and Brower 2009). Parsimony minimizes the total number of character changes, whereas the other approaches utilize models of evolution to map out the most likely character change scenarios. Brusatte et al. (2008a) used parsimony, which is the most widespread, most favored, and least assumption-ridden method for paleontological and other morphological phylogenetic studies.

However, when characters have evolved multiple times or have been lost, parsimony delivers many possible optimizations. Therefore, Brusatte et al. (2008a) used the two typical parsimony mapping techniques: accelerated (ACCTRAN) and delayed (DELTRAN) transformation optimizations (Swofford and Maddison 1987). When there is homoplasy (reversals or convergence), accelerated optimization favors losses over convergence, and therefore places the moment of character change toward the root (base) of the tree. Delayed optimization, on the contrary, favors convergence over losses and places the change closer to the tips (terminal taxa) of the tree. It is important to remember that both optimizations encompass

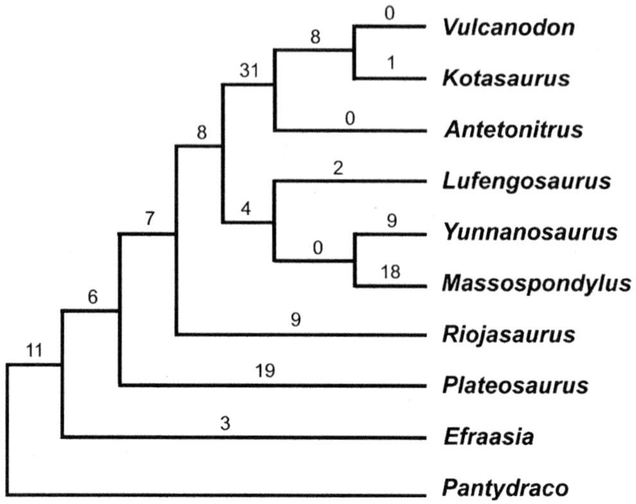

Fig. 4.2 A portion of the phylogeny utilized by Brusatte et al. (2008a), with additional Early Jurassic taxa from Brusatte et al. (2008b), with the number of raw character changes per branch indicated. These changes were determined by optimizing (under accelerated transformation) the database of morphological characters (Fig. 4.1; Table 4.1) onto the phylogeny

the same number of *total* changes, because parsimony minimizes the total number of changes, but only differ in the exactly where these moments of change are placed on the tree. In other words, accelerated and delayed optimizations will place the character changes on different branches. It is thus important to consider both approaches (and possibly others) when calculating rates.

Optimization of characters gives a total number of changes for each branch of the tree. This is the "amount" of evolution per branch. An example from the Brusatte et al. (2008a) analysis is given in Fig. 4.2. Here, the number of characters changing per branch is indicated, as optimized by accelerated transformation. Practically, phylogenetic software such as PAUP* or TNT is needed to optimize characters onto a tree, as this is too complex to do by hand for large datasets. These programs only require that the user input a tree and a dataset of characters, and the optimization is done quickly and automatically.

Determining the Number of Comparable Character Changes per Branch

However, taking just the raw number of characters changing per branch may be misleading. Perhaps one terminal taxon is only known from a skull, and therefore cannot be assessed for any of the postcranial features in the morphological dataset. As a result, it is impossible to know how many postcranial features may have changed on the branch leading to that taxon, meaning that the raw number of changes on that branch is likely to be an underestimate. Because of this problem it is necessary to correct the raw numbers to account for missing data.

The most straightforward correction was spelled out by Wagner (1997) and later utilized by Ruta et al. (2006) and Brusatte et al. (2008a). These authors measured what is called the patristic dissimilarity of each branch, which is defined as the raw number of changes per branch divided by the number of "comparable characters." Comparable characters are those that can be positively assessed in the taxa on both ends of the branch in question. For a terminal branch, one linking an observed taxon to its hypothetical ancestor, the number of comparable characters equals the total number of characters in the morphological dataset minus those characters scored as uncertain ("?") for the observed taxon. For all internal branches, the number of comparable characters is the same as the number of characters in the morphological dataset, because reconstructed ancestors, which form both ends of an internal branch, are given a positive score (i.e., not a "?") for each character. An example of patristic dissimilarity calculations for the Brusatte et al. (2008a) dataset is provided in Fig. 4.3.

Practically, these calculations can be done in Microsoft Excel or other basic spreadsheet packages. The user should simply copy the table of total changes per

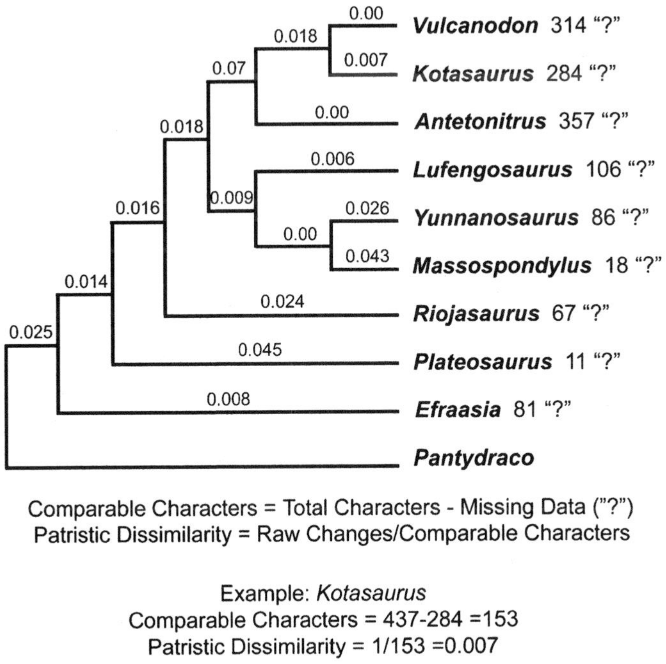

Fig. 4.3 The portion of the Brusatte et al. (2008a) phylogeny shown in Fig. 4.2, with patristic dissimilarity values listed for each branch. These values are calculated by the equation shown in the figure. The number of missing characters is denoted next to each terminal taxon on the right

Table 4.2 Spreadsheet denoting various measures necessary for evolutionary rates calculations, based on the dataset of Brusatte et al. (2008a), with the addition of Early Jurassic taxa (as employed in Brusatte et al. 2008b)

Branch to	Character changes	Missing data	Comp. characters	Patristic dissim.	Time duration	Rate
Vulcanodon	0	314	123	0.00	5.825	0.00
Kotasaurus	1	284	153	0.007	10.45	0.0007
Vulcanodon + *Kotasaurus*	8	0	437	0.018	5.825	0.003
Antetonitrus	0	357	80	0.00	2.03	0.00
A + V + K	31	0	437	0.07	2.03	0.03
Yunnanosaurus	9	86	351	0.026	2.53	0.01
Massospondylus	18	18	419	0.043	1.55	0.028
Yunnan + Masso	0	0	437	0.00	2.53	0.00
Lufengosaurus	2	106	331	0.006	2.53	0.002
Y + M + L	4	0	437	0.009	2.53	0.0036
A + V + K + Y + M + L	8	0	437	0.018	2.03	0.009
Riojasaurus	9	67	370	0.024	3.05	0.008
A + V + K + Y + M + L + R	7	0	437	0.016	3.05	0.005
Plateosaurus	19	11	426	0.045	1.81	0.025
A + V + K + Y + M + L + R + P	6	0	437	0.014	1.81	0.008
Efrassia	3	81	356	0.008	1.81	0.004
A + V + K + Y + M + L + R + P + E	11	0	437	0.025	1.81	0.014

These are the same taxa depicted in the figures. It is useful to construct this spreadsheet in Excel or another program, allowing for automated calculations when it is necessary to subtract or divide values across multiple cells

branch from the phylogenetic software output and then manually add a column denoting the comparable characters per branch (manually calculated as total characters-missing data for terminal branches and total characters for internal branches, as outlined above). From here, the total change value must be divided by the comparable character value, which can be automated by Excel or other programs. An example of a spreadsheet is presented in Table 4.2. If researchers are more familiar with computer programming, more intensive statistical packages such as R can also conduct these calculations.

Determining the Time Duration of Each Branch

Because rate is defined as changes divided by time, the patristic dissimilarity value must be divided by the duration of the branch in question. In short, it is necessary to know over what length of time the morphological changes on each branch took place. Determining the time duration of each branch on the tree can be tricky, and is open to various sources of error. For fossil taxa, calculation of these durations must rely on the absolute ages of the terminal taxa only, but for extant taxa branch durations may also be estimated by molecular sequence divergence (i.e., molecular clocks).

For paleontological datasets like that presented by Brusatte et al. (2008a), the absolute ages of the terminals are used to pull sister taxa back in time via ghost range extension (Norell 1992). For instance, if one member of a sister taxon pair (taxon A) is 10 million years old and the other (taxon B) is 5 million years old, the branch leading to taxon B would be given a 5-million year duration, because by definition this lineage must have been present when the sister taxon was present. This raises an immediate question, however: what is the duration of the branch leading to taxon A? This hits at the crux of the problem: for each pair of sister lineages, only one lineage can be extended back in time via ghost ranges. The other lineage will have a time duration of zero, and this is a fatal problem for rates analysis because time is in the denominator of the rate equation. One way around this problem is to "borrow" time from other branches. The sister clade of taxon A and taxon B has a lineage at its base, and this lineage may be extended back in time by reference to its outgroup (but only if the outgroup is older). When this time duration is calculated, it can be divided between the branch leading to taxon A and the basal branch leading to the common ancestor of taxa A and B. In essence, the time durations of branches that cannot be calculated by direct ghost range extension are calculated by sharing time from preceding branches (those more basal on the cladogram). This method was first outlined by Ruta et al. (2006) and later used, in a slightly modified form, by Brusatte et al. (2008a). An example of this method with hypothetical taxa is provided in Fig. 4.4, and an example from the Brusatte et al. (2008a) dataset is shown in Fig. 4.5.

Brusatte et al. (2008a) divided time between "shared" branches equally. In other words, if there was 20 million years of time shared between two branches (branches X and Y), then each branch was assigned a 10-million year duration. Ruta et al. (2006) outlined a slightly different method, which may be useful in many cases. They calculated the amount of time assigned to each branch to be proportional to the number of morphological characters that are estimated to have changed on that branch. In the above example, let us say that six characters changed on branch X and four characters on branch Y. The Ruta et al. (2006) method would assign an age of 12 million years to branch X and 8 million years to branch Y. These numbers are calculated because, out of the 20-million year total extension, branch X receives 60% of the duration (6/10 of the characters changed on that branch) and branch Y receives 40% of the duration (4/10 characters changed on that branch). The Brusatte et al. (2008a) method is more neutral, but the Ruta et al. (2006) method biases the statistical tests (see below) in favor of the null hypothesis of equal rates across branches or time bins. Therefore, it may be preferred when researchers want to make it more likely that the data conform to the null hypothesis, thus demanding more stringent criteria to reject the null.

Practically, it is possible to do these calculations by hand or with automated computer software. When dating branches by hand, it is easiest to draw out the tree, as shown in Figs. 4.4 and 4.5, with absolute ages next to each terminal taxon. From here the user can work from the tips of the tree towards the root, extending branches by virtue of sister taxon comparisons and dividing time between branches with a calculator. Alternatively, for those well versed in statistical software, Graeme

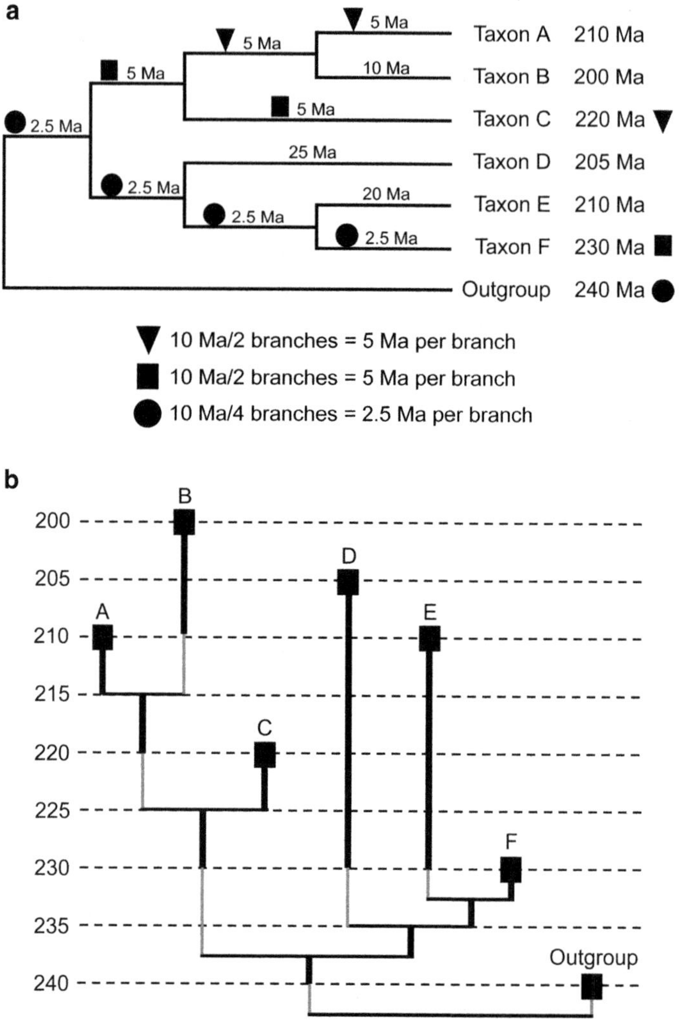

Fig. 4.4 A hypothetical example illustrating how to date branch durations on a phylogeny using the method employed in this paper (see Ruta et al. 2006; Brusatte et al. 2008a). In (**a**), the absolute ages of the terminal taxa are listed to the right. Branches are dated by extending them back in time by reference to their sister taxon. As outlined in the text, only one sister taxon per pair can be extended this way, and the other must be extended by sharing time with preceding branches that are extended relative to more distant outgroups. Branches that share time are denoted by a shape (*triangle*, *circle*, or *square*), and the sister taxon that extends these series of branches is marked by the corresponding symbol. In (**b**), these extensions are shown to scale relative to a timescale on the left. The *thick black bars* are the actual branch durations, whereas *thin gray bars* are not counted as part of the duration (they are used to show the relationships more clearly, but the *thick black bars* are the proper durations because they extend to the age of the sister taxon)

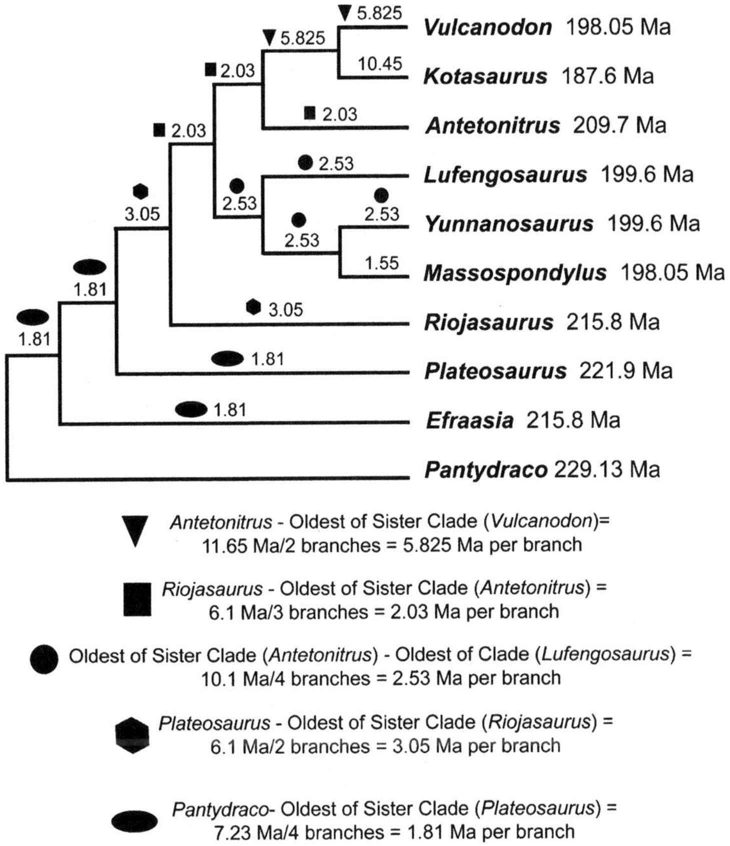

Fig. 4.5 The portion of the Brusatte et al. (2008a) phylogeny shown in Figs. 4.2 and 4.3, with time durations listed for each branch. These values are calculated by the sister taxon extension method outlined in the text and Fig. 4.4. Branches that share time are denoted by a shape (*triangle*, *circle*, or *square*), and these extensions are explained in the legend at bottom

Lloyd (Natural History Museum, London, http://www.graemetlloyd.com) provides a free function that can date trees in the program R (http://cran.r-project.org/).

Calculating the Rate of Each Branch

At this stage, calculating the rate of each branch is merely a matter of dividing the patristic dissimilarity by the time duration (Fig. 4.6). These calculations can be done in Excel or another spreadsheet program, and it is easiest to do all calculations in a single spreadsheet file. Therefore, it is recommended that the researcher add a column for time duration next to the column for patristic dissimilarity in the spreadsheet discussed in Sect. 3.2 (Table 4.2). Then, the software program can

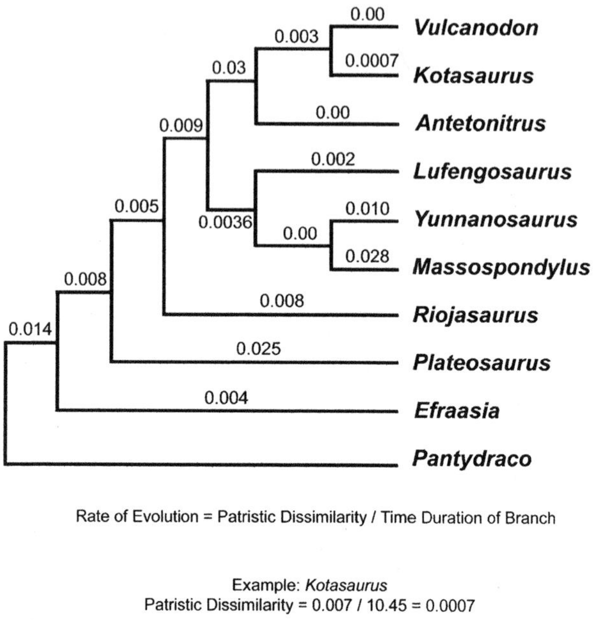

Fig. 4.6 The portion of the Brusatte et al. (2008a) phylogeny shown in Figs. 4.2, 4.3, and 4.5, with morphological rates listed for each branch. These values are calculated by dividing the patristic dissimilarity of each branch (see Fig. 4.3) by the time duration of that branch (see Fig. 4.5)

automatically divide the patristic dissimilarity value for each branch by the time duration, giving a final column of morphological rate for each branch.

Binning Rates to Address Macroevolutionary Questions

With a rate value now pinned to each branch it is up to the researcher to decide how to marshal these data in additional quantitative analyses. Brusatte et al. (2008a) were interested in the major patterns of morphological rate over time and across the phylogeny. Therefore, they binned rates according to time (placing each rate value in the time bin of the first observed member of the branch in question) and clade. They then plotted these values, resulting in a curve of archosaur rates over time (akin to a curve of diversity over time) and pairwise comparisons of various major archosaur subgroups (Fig. 4.7). Researchers working on other groups, and interested in other questions, may want to bin rates in different ways. Perhaps a researcher would like to know if organisms from a certain geographic location evolved faster or slower than taxa from a different locale. In this case, rates could be binned by geographic occurrence. Maybe another researcher is interested in knowing whether genera with numerous species evolve faster than those with few species. Here, speciose genera could be binned relative to depauperate genera.

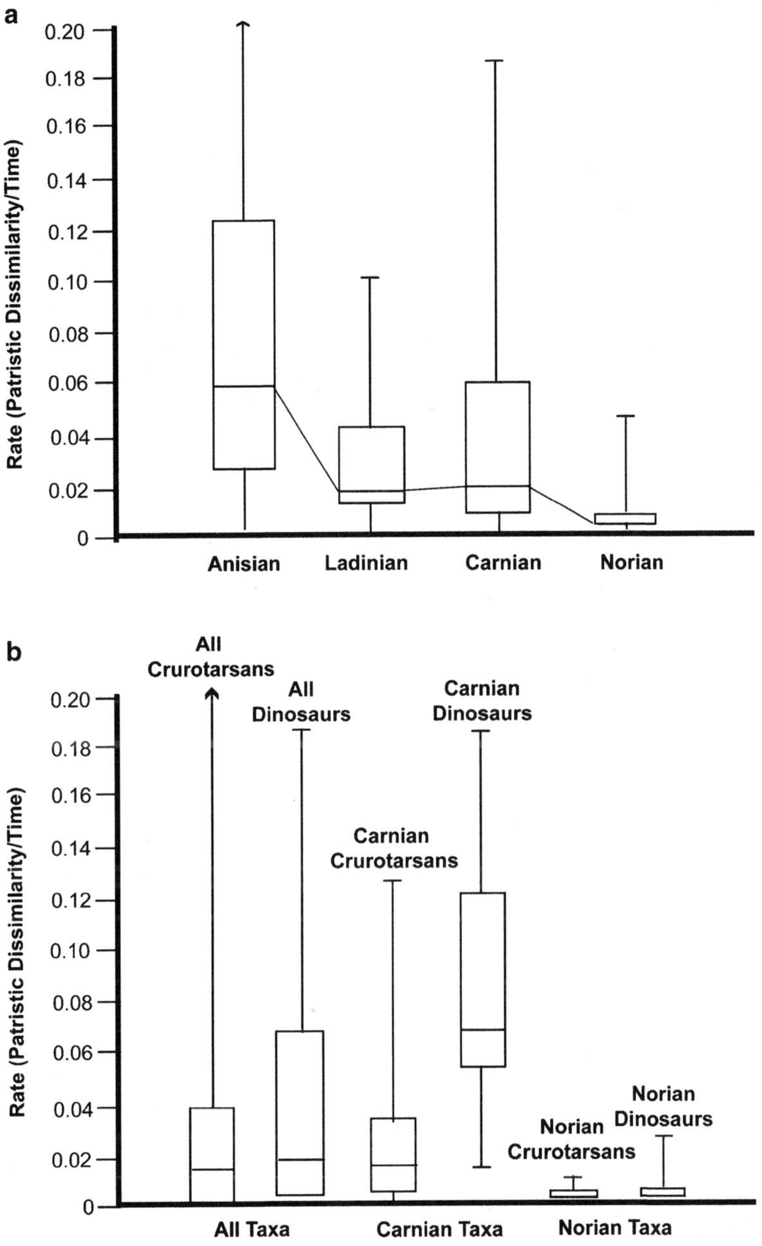

Fig. 4.7 Two examples of how morphological rates can be binned and compared to gain insight on evolutionary patterns and processes, modified from Brusatte et al. (2008a). (**a**) Shows morphological rates in all archosaurs across the Triassic, whereas (**b**) compares the rates in dinosaurs and crurotarsan (crocodile-line) archosaurs during the Triassic as a whole, the Carnian, and the Norian. The plots depict the distributions of real rate data for each bin. The boxes represent the 25th–75th percentiles, with the horizontal line depicting the mean. The whiskers bracket the 5th–95th percentiles. Whether differences between individual bins are significant or not is assessed by the Mann–Whitney U test (see Tables 4.3 and 4.4). See text for summary and interpretation of results

The possibilities for rate comparisons are limitless once the basic data – a single rate calculated for each branch of the tree – are available.

Statistical Tests to Determine Rate Heterogeneity

Once the researcher decides how to bin rate values, it is necessary to determine statistical significance: are the differences in rates between time periods or clades truly significant or simply marginal? Various statistical tests are possible, but the most useful for comparisons between different bins is the Mann–Whitney U test (as used by Wagner 1997; Brusatte et al. 2008a). This test assesses the probability of whether two samples – in this case, rates binned by time, clade, etc. – come from the same distribution (Wilcoxon 1945; Mann and Whitney 1947). Each bin of rates has a spread of values: some high, some low, many in between. This spread comprises a distribution. Comparisons of distributions are more powerful than simple comparisons between average or median values, because they take into account the entire spread of rate observations rather than a simple representative value. Operationally, the Mann–Whitney test begins with a null hypothesis that two rate distributions are equal (in other words, drawn from the same larger distribution), and a significant comparison is one that violates this null. Importantly, the Mann–Whitney test is non-parametric; that is, it does not assume that the rates in each bin are normally distributed, an idealized scenario which is probably not the case with most real-world rate data. Practically, the analyses must be carried out using a software package. A recommended program is PAST, a statistical package designed specifically for paleontologists (Hammer et al. 2001).

Caveats for Rate Analysis

Following the above guidelines, a researcher can easily calculate a rate of change for each branch of the phylogeny and bin these to address various evolutionary questions. Before interpreting the results, however, it is important for the researcher to understand possible sources of error and bias.

First, rate analysis should only be undertaken on a species-level phylogeny, because it is crucial that all terminal taxa are of equivalent rank. It is also ideal if this species-level phylogeny is complete, or in other words, includes all known taxa. If taxa are missing, then it is difficult to properly measure the number of character changes on a branch. In some cases, this will result in an overestimate: a "long" branch with numerous character changes that would be broken into many smaller branches if other taxa were sampled. In other cases, however, missing taxa may result in an underestimate: perhaps having additional taxa to observe and measure would reveal additional character changes. Unfortunately, in practice it is often difficult or impossible to use a complete species-level phylogeny. If so, species must still be used as terminal taxa, but a consistent representation strategy using exemplars should be employed. Brusatte et al. (2008a) could not sample every Triassic archosaur species, but consistently represented each major archosaur subgroup with a spread of exemplar species that encompassed the major body plans and morphologies within each group. This way, problems with missing lineages are as evenly spread across the phylogeny as possible, not concentrated in a particular group.

In sum, it is important that sampling is even across the tree: no one part of the tree should be better sampled than another. If sampling is incomplete, then biases must be evenly spread around the tree. These biases may be due to the calculated decision of the researcher not to sample certain lineages (exemplar strategy, see above), as well as other factors that cause lineages to be unknown or unsampled (extinction, non-preservation). In some cases it is reasonable to assume that extinction and non-preservation are spread evenly across the phylogeny (i.e., that there is no reason to think that certain clades have more unsampled lineages than others). However, perhaps some time periods are better sampled than others, or some groups have a poorer fossil record than others because their habitat or lifestyle is less amenable to preservation. If so, this violates the condition of even sampling across the tree. It is difficult to correct for such biases in the fossil record, as is sometimes possible with more simplistic measures such as raw taxonomic diversity. However, the researcher should be open about such biases, and keep them in mind when interpreting results.

Finally, researchers should also keep in mind the major sources of error inherent in morphological rates analysis: phylogenetic topology and absolute age durations. The rates calculated for each branch, and therefore the spread of rates for each time interval or clade, depend intimately on the phylogeny used and the dates assigned to the terminals. If possible, researchers should experiment with a sample of different phylogenies, and if the same results are found then additional confidence is earned. Similarly, it is recommended that researchers do not use a single age for each terminal, but rather employ a randomization approach that considers many possible ages drawn from the finest age resolution in the fossil record (Pol and Norell 2006).

Interpretation of Morphological Rates

The final stage of rates analysis is interpretation. By this stage, the patterns – the waxing and waning of rates over time, differences in rates between major clades, etc. – are established and tested statistically. What do these patterns reveal about the evolutionary history of the organisms in question or evolutionary processes?

Brusatte et al. (2008a) focused on the evolution of archosaurian reptiles during the Triassic, including dinosaurs, pterosaurs (flying reptiles), and early relatives of the crocodiles (crurotarsans). As outlined above, they binned rates by both time and clade in order to assess whether archosaur rates changed over time and whether certain clades had higher or lower rates than others. In short, they aimed to establish

Table 4.3 Results of the statistical test (Mann–Whitney U test) for comparisons of archosaur morphological rates across the Triassic

	Anisian	Ladinian	Carnian	Norian
Anisian	X	U = 143, $p = 0.0213$ *SIG*	U = 309, $p = 0.0423$ *SIG*	U = 106, $p = 0.00003$ *SIG*
Ladinian	X	X	U = 455, $p = 0.5997$	U = 100, $p = 0.000005$ *SIG*
Carnian	X	X	X	U = 223, $p = 0.0000006$ *SIG*
Norian	X	X	X	X

Each test is a pairwise comparison of one time bin against another bin
"U" refers to the test statistic and p to the probability value, and "sig" denotes a significant comparison (i.e., one that violates the null hypothesis of equal rate distributions between bins, and thus supports the finding that one bin has significantly higher rates than another)
The general pattern is that archosaur rates were highest early in the Triassic and decrease over time, which is borne out by the statistical results

rate patterns that could then be interpreted to give insights on the grand narrative of early archosaur evolution, especially the origin and radiation of dinosaurs.

First, Brusatte et al. (2008a) found that, as a whole, archosaur rates were significantly highest early in the clade's history and proceeded to decrease over time (Fig. 4.7a; Table 4.3). This pattern is indicative of a rapid initial radiation of archosaurs, consistent with predictions from macroevolutionary theory that rates of character change are elevated during major biodiversification events (e.g., Valentine 1980; Schluter 2000; Gould 2002). Taking things a step further, Brusatte et al. (2008a) interpreted the pattern as reflecting a great burst of character evolution during the first few million years of archosaur history, perhaps as a direct consequence of vacant ecospace that was available after the Permo–Triassic mass extinction. Interestingly, archosaurs experienced their significant increases in taxonomic diversity, absolute faunal abundance, and morphological disparity (variety of morphological features instead of rates of change) in the Late Triassic, tens of millions of years after their peak morphological rates (Benton 1983; Brusatte et al. 2008a, b). This is prime evidence that different aspects of the archosaur radiation were decoupled: the first stage of archosaur history witnessed a rapid pace of character change, and only when rates slowed down, and the major subclades were established, did archosaurs truly expand into globally distributed, diverse, and abundant components of terrestrial faunas.

Second, Brusatte et al. (2008a) explicitly compared the morphological rates of dinosaurs and crocodile-line archosaurs (crurotarsans). Accumulating evidence has revealed that these two groups were eerily convergent on each other during the Late Triassic, and for tens of millions of years they lived alongside each other and likely inhabited similar niches (Parker et al. 2005; Nesbitt and Norell 2006; Nesbitt 2007). One longstanding narrative in vertebrate paleontology is that dinosaurs gradually outcompeted other reptile groups, including their close crurotarsan cousins, over millions of years during the Late Triassic (e.g., Bakker 1971; Charig 1984). However, the emerging evidence that dinosaurs and crurotarsans were so similar to each other calls this hypothesis into question. Brusatte et al. (2008a) found that dinosaurs and crurotarsans were essentially evolving at similar rates during the Late

Table 4.4 Results of the statistical test (Mann–Whitney U test) for comparisons of dinosaur and crurotarsan (crocodile-line) archosaur morphological rates across the Triassic

Time comparison	U	*p*
Entire Triassic	192	0.4812
Carnian	40	0.0003967 *SIG*
Norian	38	0.1003

Each test is a pairwise comparison of dinosaur rates vs. crurotarsan rates, and comparisons are made for all Triassic taxa as a whole as well as those that lived within individual time bins (Carnian, Norian)
"U" refers to the test statistic and *p* to the probability value, and "sig" denotes a significant comparison (see Table 4.3 for explanation)
The general pattern is that dinosaurs and crurotarsans were evolving at statistically indistinguishable rates, except that there is statistical evidence that Carnian dinosaurs were evolving faster than Carnian crurotarsans

Triassic (Fig. 4.7b; Table 4.4). In other words, there are no significant differences between the rates of morphological evolution in dinosaurs and their supposed competitors. There is evidence that dinosaurs were evolving faster than crurotarsans during one stage of the Late Triassic, the Carnian, but otherwise there are no statistically robust differences between the groups. Brusatte et al. (2008a) interpreted this pattern as a lack of evidence for the traditional notion of dinosaurs outcompeting crurotarsans. Of course, these results do not disprove or falsify competition, but in revealing no meaningful differences between dinosaurs and crurotarsans provide another line of evidence that the groups had similar evolutionary and ecological trajectories during the Triassic.

These are only two examples of the types of comparisons and interpretations that can be made in an evolutionary rates analysis. Other possibilities abound. Quantifying rates may be a valuable test of whether a radiation was adaptive, as high rates immediately after a clade colonizes new ecospace would be congruent with such a scenario (Schluter 2000; Gavrilets and Losos 2009). The classic view of an adaptive radiation also holds that morphological change and lineage divergence (speciation) are associated with each other. Such an association may also be predicted in cases of punctuated equilibrium, when long periods of stasis in a lineage are interrupted by rapid episodes of lineage splitting and morphological change (Gould and Eldredge 1977; Stanley 1979; Gould 2002). Therefore, researchers may be interested in testing whether rates of speciation (diversification) and morphological change are congruent during the history of a group, to gauge whether adaptive radiation or punctuated change took place (Adams et al. 2009). Other workers may be interested in testing whether certain key adaptations are associated with clades that undergo abnormally high or low amounts of morphological change. These comparisons are not often made in the literature, as most studies focus on correlating key adaptations with especially diverse clades (e.g., Chan and Moore 2002; Sims and McConway 2003; McConway and Sims 2004; Moore and Donoghue 2009), but hold great potential for future exploration. Furthermore, researchers may be curious about dynamics of faunal replacement over time, and comparing trends in the evolutionary rates in different clades, especially competitors or coexisting groups, may be paramount.

As is evident, quantifying and comparing morphological rates of evolution can be useful in a wide variety of paleontological and evolutionary studies. Researchers have, for many decades, focused more heavily on lineage diversity relative to morphological change. However, as new fossils are discovered and new methods make it easier to quantify and study morphology in an explicit framework, studies of morphological evolution are becoming more prevalent (e.g., Wills et al. 1994; Foote 1997; Wagner 1997; Ciampaglio et al. 2001; Collar et al. 2005; O'Meara et al. 2006; Ruta et al. 2006; Erwin 2007; Sidlauskas 2007; Brusatte et al. 2008a, b; Pinto et al. 2008; Adams et al. 2009; Cooper and Purvis 2009; Ruta 2009). Morphological rates analysis is straightforward but powerful, and is still a mostly untapped area for future work. Most clades have yet to be studied in this sense, and such studies promise to become a major component of evolutionary biology research in years to come.

Acknowledgements I thank Ashraf Elewa for his invitation to contribute to this volume, Prof. Richard Reyment and Graeme Lloyd for their reviews, and my collaborators on various projects relating to morphological rates (Graeme Lloyd, Steve Wang, Mike Benton, and Marcello Ruta). For discussions regarding rates and Triassic archosaurs I think many people, most importantly Richard Butler, Michael Coates, Philip Donoghue, Mike Foote, Randall Irmis, Michael LaBarbera, Sterling Nesbitt, Grzegorz Niedźwiedzki, Mark Norell, Paul Olsen, Paul Sereno, Mark Siddall, and Ward Wheeler. My work on morphological rates and Triassic archosaurs has been funded by a Marshall Scholarship for study in the UK (University of Bristol) and an NSF Graduate Research Fellowship (Columbia University), and specimen visits pertinent to this project were funded by the Paleontological Society, Jurassic Foundation, SYNTHESYS, and Bob Savage Memorial Fund (University of Bristol).

References

Adams DC, Berns CM, Kozak KH, Wiens JJ (2009) Are rates of species diversification correlated with rates of morphological evolution? Proceedings of the Royal Society, Series B 276: 2729–2738.

Bakker RT (1971) Dinosaur physiology and the origin of mammals. Evolution 25: 636–658.

Benton MJ (1983) Dinosaur success in the Triassic: a noncompetitive ecological model. Quarterly Review of Biology 58: 29–55.

Bookstein F (1991) Morphometric tools for landmark data: geometry and biology. Cambridge University Press, Cambridge.

Brusatte SL (2007) The higher-level phylogeny of Archosauria (Tetrapoda: Diapsida). Unpublished MSc thesis, University of Bristol: Bristol.

Brusatte SL, Benton MJ, Desojo JB, Langer MC (2010) The higher-level phylogeny of Archosauria (Tetrapoda: Diapsida). Journal of Systematic Palaeontology 8: 3–47.

Brusatte SL, Benton MJ, Ruta M, Lloyd GT (2008a) Superiority, competition, and opportunism in the evolutionary radiation of dinosaurs. Science 321: 1485–1488.

Brusatte SL, Benton MJ, Ruta M, Lloyd GT (2008b) The first 50 myr of dinosaur evolution: macroevolutionary pattern and morphological disparity. Biology Letters 4: 733–736.

Chan KMA, Moore BR (2002) Whole-tree methods for detecting differential diversification rates. Systematic Biology 51: 855–865.

Charig AJ (1984) Competition between therapsids and archosaurs during the Triassic Period: a review and synthesis of current theories. Symposium of the Zoological Society of London 52: 597–628.

Ciampaglio CN, Kemp M, McShea DW (2001) Detecting changes in morphospace occupation patterns in the fossil record: characterization and analysis of measures of disparity. Paleobiology 27: 695–715.

Cloutier R (1991) Patterns, trends, and rates of evolution within the Actinistia. Environmental Biology of Fishes 32: 23–58.

Collar DC, Near TJ, Wainwright PC (2005) Comparative analysis of morphological diversity: does disparity accumulate at the same rate in two lineages of centrarchid fishes? Evolution 59: 1783–1794.

Cooper N, Purvis A (2009) What factors shape rates of phenotypic evolution? A comparative study of cranial morphology of four mammalian clades. Journal of Evolutionary Biology 22: 1024–1035.

Derstler KL (1982) Estimating the rate of morphological change in fossil groups. Proceedings of the Third North American Paleontological Convention 1: 131–136.

Elewa A (ed) (2004) Morphometrics—Applications in Biology and Paleontology. Springer-Verlag, Heidelberg, Germany.

Erwin DH (2007) Disparity: morphological pattern and developmental context. Palaeontology 50: 57–73.

Felsenstein J (2004) Inferring Phylogenies. Sinauer Associates, Inc., Sunderland, Massachusetts.

Foote M (1997) The evolution of morphological diversity. Annual Review of Ecology and Systematics 28: 129–152.

Forey PL (1988) Golden jubilee for the coelacanth Latimeria chalumnae. Nature 336: 727–732.

Gavrilets S, Losos JB (2009) Adaptive radiation: contrasting theory with data. Science 323: 732–737.

Goloboff PA, Farris JS, Nixon KC (2008) TNT, a free program for phylogenetic analysis. Cladistics 24: 774–786.

Gould SJ (2002) The Structure of Evolutionary Theory. Harvard University Press, Cambridge, Massachusetts.

Gould SJ, Eldredge N (1977) Punctuated equilibria: the tempo and mode of evolution reconsidered. *Paleobiology* 3: 115–151.

Hammer Ø, Harper DAT, Ryan PD (2001) PAST: Paleontological statistics software package for education and data analysis. Palaeontologia Electronica 4(4): 1–9.

Huelsenbeck JP, Ronquist F, Nielsen R, Bollback JP (2001) Bayesian inference of phylogeny and its impact on evolutionary biology. *Science* 294: 2310–2314.

Jernvall J, Hunter JP, Fortelius M (1996) Molar tooth diversity, disparity, and ecology in Cenozoic ungulate radiations. Science 274: 1489–1492.

Kitching IJ, Forey PL, Humphries CH, Williams DM (1998) Cladistics: The Theory and Practice of Parsimony Analysis. The Systematics Association, London.

Mann HB, Whitney DR (1947) On a test of whether one of two random variables is stochastically larger than the other. Annals of Mathematical Statistics 18: 50–60.

McConway KJ, Sims HJ (2004) A likelihood-based method for testing for nonstochastic variation of diversification rates in phylogenies. Evolution 58: 12–23.

McGowan AJ, Dyke GJ (2007) A morphospace-based test for competitive exclusion among flying vertebrates: did birds, bats and pterosaurs get in each other's space? Journal of Evolutionary Biology 20: 1230–1236.

Moore BR, Donoghue MJ (2009) A Bayesian approach for evaluating the impact of historical events on rates of diversification. Proceedings of the National Academy of Sciences (USA) 106: 4307–4312.

Nesbitt SJ (2007) The anatomy of *Effigia okeeffeae* (Archosauria, Suchia), theropod- like convergence, and the distribution of related taxa. Bulletin of the American Museum of Natural History 302: 1–84.

Nesbitt SJ, Norell MA (2006) Extreme convergence in the body plans of an early suchian (Archosauria) and ornithomimid dinosaurs (Theropoda). Proceedings of the Royal Society of London, Series B 273: 1045–1048.

Norell MA (1992). Taxic origin and temporal diversity: the effect of phylogeny. In: Novacek MJ, Wheeler QD (eds), Extinction and Phylogeny. Columbia University Press, New York, p. 88–118.

O'Meara BC, Ané C, Sanderson MJ, Wainwright PC (2006) Testing for different rates of continuous trait evolution using likelihood. Evolution 60: 922–933.

Parker WG, Irmis RB, Nesbitt SJ, Martz JW, Browne LS (2005) The Late Triassic pseudosuchian *Revueltosaurus callenderi* and its implications for the diversity of early ornithischian dinosaurs. Proceedings of the Royal Society of London, Series B 272: 963–969.

Pinto G, Mahler DL, Harmon LJ, Losos JB (2008) Testing the island effect in adaptive radiation: rates and patterns of morphological diversification in Caribbean and mainland Anolis lizards. Proceedings of the Royal Society of London, Series B 275: 2749–2757.

Pol D, Norell MA (2006) Uncertainty in the age of fossils and the stratigraphic fit to phylogenies. Systematic Biology 55: 512–521.

Ruta M (2009) Patterns of morphological evolution in major groups of Palaeozoic Temnospondyli (Amphibia: Tetrapoda). Special Papers in Palaeontology 81: 91–120.

Ruta M, Wagner PJ, Coates MI (2006) Evolutionary patterns in early tetrapods. I. Rapid initial diversification followed by decrease in rates of character change. Proceedings of the Royal Society of London Series B 273: 2107–2111.

Schluter D (2000) The Ecology of Adaptive Radiation. Oxford University Press, Oxford.

Schuh RT, Brower AVZ (2009) Biological Systematics: Principles and Applications (2nd Edition). Cornell University Press, Ithaca, NY.

Sidlauskas BL (2007) Testing for unequal rates of morphological diversification in the absence of a detailed phylogeny: a case study from characiform fishes. Evolution 61: 299–316.

Sims HJ, McConway KJ (2003) Nonstochastic variation of species-level diversification rates within angiosperms. Evolution 57: 460–479.

Stanley SM (1979) Macroevolution: patterns and process. Freeman: New York.

Swofford DL (2000) *PAUP*: Phylogenetic Analysis Using Parsimony* (*and other methods), Version 4.10b. Released by the author.

Swofford DL, Maddison WP (1987) Reconstructing ancestral character states under Wagner parsimony. Mathematical Biosciences 87: 199–229.

Valentine JW (1980) Determinants of diversity in higher taxonomic categories. Paleobiology 6: 444–450.

Wagner PJ (1997) Patterns of morphologic diversification among the Rostroconchia. Paleobiology 23: 115–150.

Walker JD, Geissman JW (2009) 2009 GSA Geologic Time Scale. GSA Today April/May 2009: 60–61.

Westoll TS (1949) On the evolution of the Dipnoi. In: Jepsen GL, Simpson GG, Mayr E (eds), Genetics, Paleontology and Evolution. Princeton University Press, Princeton, p. 121–184.

Wilcoxon, F (1945) Individual comparisons by ranking methods. Biometrics Bulletin 1: 80–83.

Wills MA, Briggs DEG, Fortey RA (1994) Disparity as an evolutionary index: a comparison of Cambrian and Recent arthropods. Paleobiology 20: 93–131.

Zelditch ML, Swiderski DL, Sheets HD, Fink WL (2004) Geometric morphometrics for biologists: a primer. Elsevier Academic Press, New York.

Chapter 5
Computational Model of Growth and Development in Paleozoic Echinoids

Louis G. Zachos and James Sprinkle

Introduction

The term *theoretical morphology* is a composite term, with a central theme of morphology. The *theoretical* part of the term implies some degree of removal from the organism. By this distancing of morphology from the organism, theoretical morphology tries to explain morphology or aspects of morphology in terms of core principles of geometry, function, inheritance, growth, and development (McGhee 1999). Theoretical morphology can be restricted to the simulation of biological form via "programs" of morphogenesis or growth (Reif and Weishampel 1991). These can be actual computer programs, purely mathematical or analytical models, or physical models. Two basic kinds of morphologic models can be differentiated (Konarzewski et al. 1998). The traditional model, meant to provide an empirical description of growth, has the goal of detecting patterns of growth among organisms. These models emphasize generality and simplicity. The traditional approach is descriptive and mostly falls within the domain of biometrics. This approach is exemplified by the morphometric paradigm (Bookstein 1996). The second approach is the simulation of the mechanisms that cause morphogenesis. This dichotomy of meaning has been reiterated as either simulation of some aspect of form with a simplified set of parameters or simulation of the morphogenetic processes themselves (McGhee 1999).

The current trend in theoretical morphology is to exclude the descriptive and strictly morphometric approach and concentrate on modeling within the architecture of evolutionary and developmental biology, or "evo-devo" (Hall 1992). Even in this restricted sense, the field is difficult to constrain. Modeling of developmental processes can involve simulations of developmental pathways at the molecular level

L.G. Zachos (✉)
Non-vertebrate Paleontology Laboratory, Texas Natural Science Center (Texas Memorial Museum), 10100 Burnet Rd, Austin, TX 78758-4445, USA
e-mail: lg_zachos@alumni.utexas.net

J. Sprinkle
Department of Geological Sciences, Jackson School of Geosciences, University of Texas, Austin, TX 78712-0254, USA

A.M.T. Elewa (ed.), *Computational Paleontology*,
DOI 10.1007/978-3-642-16271-8_5,

or extrapolation to higher-level processes (Wilkins 2002). Developmental biology and computer science can be linked by a unifying theme: construction (Kumar and Bentley 2002). How are organisms constructed, and how can computer programs be constructed to simulate organismal development? This facet of theoretical morphology has its own name: computational systems biology (Kitano 2002).

The field generally described as *computational paleontology* has been focused almost exclusively on reconstruction of fossil animals and the use of imaging and computational tools to render three-dimensional images of fossils, either in their current form or as an interpretation of how they appeared when alive. In contrast, this chapter will describe a methodology more closely aligned with that of the developmental biology school of thought.

The echinoderm class Echinoidea, containing such modern animals as sea urchins and sand dollars, first appeared in the fossil record in the Late Ordovician. The abundance and overall diversity of the group was low throughout the Paleozoic, although the disparity in forms, as represented by the variable number of plate columns in the skeleton, was high. The class nearly became extinct at the end of the Paleozoic, and only one body design survived into the Triassic. All crown group echinoids invariably have 20 columns of plates in the corona, arranged as five sets of biserial ambulacral plates and five sets of biserial interambulacral plates. The stem group Paleozoic echinoids have from 15 to more than 150 columns in various configurations.

Two major growth strategies have been defined as explanation for the morphological differences between Paleozoic and post-Paleozoic echinoids (Smith 2005). The growth strategies are promoted by two distinct mechanisms of growth found in all echinoids: plate addition and plate accretion. According to this hypothesis, growth in Paleozoic echinoids was achieved almost exclusively by plate addition, whereas in many crown group echinoids growth is controlled in large part by plate accretion and remodeling. These two mechanisms of echinoid skeletal growth have been shown via a computational model to be sufficient to explain the major aspects of skeleton growth in modern, crown-group regular sea urchins (Zachos 2009). This study also showed that there are important geometric aspects to the resulting patterns indicating significant constraints on these simple mechanisms.

The proposal that growth in Paleozoic echinoids was predominated by plate addition is incomplete in that it leaves unresolved the question of how plates are added, and whether or not the process differs between the Paleozoic and modern echinoids. It also fails to define whether or not there are differences in how plate accretion proceeds between the groups. The complex patterns of plates in Paleozoic echinoids make it very difficult to deduce the processes involved in their formation directly. However, it is exactly this geometric complexity that may hold the answers to these questions, and modeling is an approach that can evaluate hypotheses regarding the growth of these animals.

Materials and Methods

Studies of the modern green sea urchin *Strongylocentrotus droebachiensis* used specimens collected in coastal Maine. Early-stage post-metamorphic specimens (imagos) were obtained from a sea urchin breeding facility at the R.J. Peacock Hatchery in

Lubec, Maine. Specimens were examined under a binocular microscope, and photographed using digital cameras and various selections of lenses and extenders. Additional imagery was obtained in some cases using a JEOL JSM-T330A Scanning Electron Microscope (SEM) on gold-coated specimens and a Phillips Environmental SEM (ESEM) on gold–palladium coated and uncoated specimens.

Recent and fossil material was examined from collections at the National Museum of Natural History (USNM), Washington, DC; Harvard University Museum of Comparative Zoology (MCZ), Cambridge, Massachusetts; Field Museum (UC), Chicago, Illinois; and Texas Natural Science Center (NPL), Austin, Texas.

The computational model was written in the programming language C++, compiled using Microsoft® Visual Studio with the Graphical User Interface (GUI) configured for the Microsoft Windows® operating systems.

Echinoid Morphology

Modern Echinoids

Echinoids are characterized by an internal, plated skeleton called the test. The test consists of individual plates, each composed of a meshwork of calcite termed stereom, composed of calcite trabeculae, and pores filled with mesodermal tissue termed stroma (Smith 1980). The skeleton is internal, covered almost everywhere in epithelium, and the individual plates are enclosed within syncytial cell membranes (Märkel et al. 1989). The plates fit together in a mosaic to create a more or less rigid framework (Ellers et al. 1998), although in many Paleozoic and some modern forms the plates are imbricate and the test flexible (Jackson 1912). The skeleton is divided into three structurally distinct parts called the corona, which makes up the major portion of the echinoid skeleton; the apical system of plates, located at the apex of the test and which usually consists of four or five genital plates and five ocular plates; and the peristomal and periproctal plates, which cover the peristomal and periproctal membranes associated with the mouth and anus. The plates of the corona are arranged in columns in a pentaradial fashion, and can be divided into the ambulacral plates, which lie in a radial position and are associated with the pores for tube feet of the water vascular system, and the interambulacral plates, which lie in an interradial position and are generally imperforate. The corona in all post-Paleozoic echinoids (with the exception of two problematic genera) is composed of ten ambulacral plate columns and ten interambulacral columns, as five sets of biserial columns, respectively.

Paleozoic Echinoids

The echinoids known from the Paleozoic are generally referred to as stem-group echinoids (Smith 2005). Nearly all are characterized by more than 20 columns of coronal plates, which can be multiple columns of ambulacral plates, interambulacral plates, or both. The sets of columns are still arranged pentaradially, and the apical systems, when they can be observed, are usually composed of five genital and five ocular plates, in the interradial and radial positions, respectively. The Paleozoic echinoids (if the problematic Bothriocidaridae are excluded) can be broadly divided into those with a semi-rigid test and all others with imbricate plates and a flexible test.

Modes of Growth

Ambulacral Plate Addition

Märkel (1981) showed that for *Eucidaris tribuloides* ambulacral plates are added from the inside of the test and interambulacral plates from the outside. This appears to be the case for all modern echinoids, although it can best be seen in regular sea urchins with large apical systems. This character suggests that there is an important distinction between how the two types of coronal plates are added. In *Psammechinus miliaris* the ocular plates first appear before metamorphosis in the larval rudiment above the forming terminal tube feet of the water vascular system, each plate later growing to surround the tube foot (Gordon 1926). In *S. droebachiensis* the ocular does not grow to surround the terminal tube foot until after metamorphosis (Fig. 5.1). This pattern of development of a tube foot and growth of skeleton to surround the tube foot is the case for the coronal ambulacral plates as well (Fig. 5.2). Initially, a single pore is developed, only later developing the stereom to separate the tube

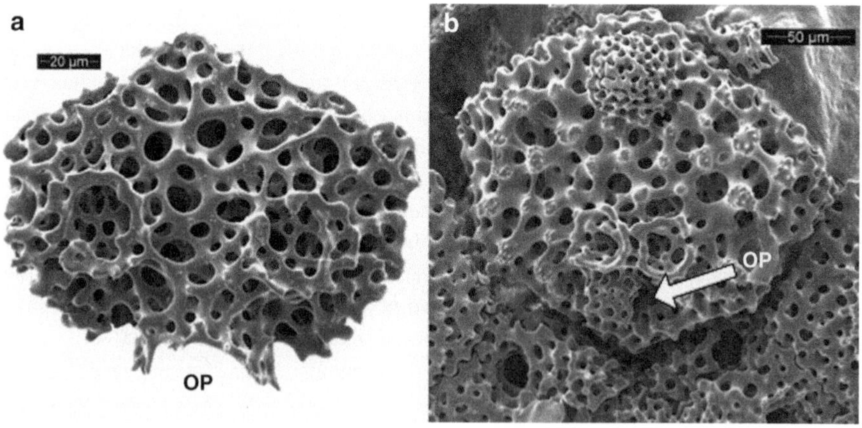

Fig. 5.1 Ocular plates in *Strongylocentrotus droebachiensis*. (**a**) Plate from imago 1 day post-metamorphosis. Note that ocular pore has not closed around terminal tube foot. (Scale bar 20 µm) (**b**) Plate from juvenile 3 months old. Ocular pore now completely encloses the terminal tube foot. *OP* ocular pore (Scale bar 50 µm)

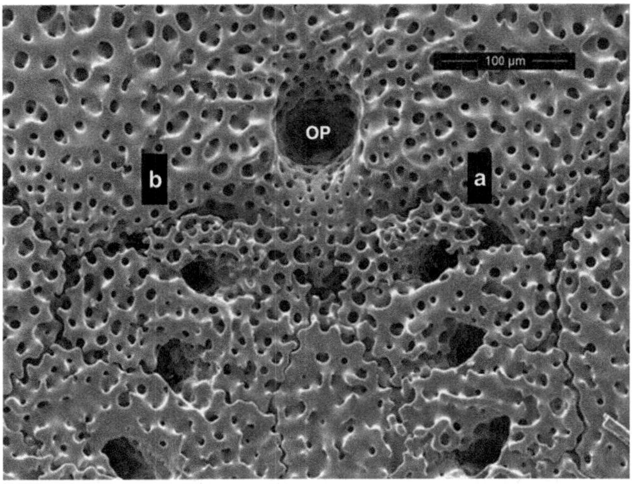

Fig. 5.2 Addition of new ambulacral plates in *Strongylocentrotus droebachiensis*, internal view showing ocular plate and proximal section of ambulacrum. The youngest plates in the ambulacrum (a and b) have not closed around their respective tube feet. *OP* ocular pore (Scale bar 100 μm)

foot pair that exits to the surface of the test. The ambulacral plates originate inside the test, and the formation of ambulacral plates appears to be directly induced by the formation of a lateral canal of the water vascular system.

The direct association between the lateral canals of the water vascular system and the ambulacral plates equates the model of ambulacral plate addition to simulation of formation of alternating left and right lateral branches from the radial water canal (Cavey and Märkel 1994). For each individual ambulacrum, the radial water canal is analogous to the central trunk of a tree, and the lateral branches to the branches of the tree. Märkel (1981) showed that the length of the lateral canals is at a minimum nearest the terminal end of the radial water canal, and progressively longer farther from the end. In its simplest form, in which the lateral branches quickly reach a maximum length which is then maintained during further growth, the model creates a simple ambulacrum of a type seen in cidaroids and other urchins with non-compound plates. If the lateral water canal length constraint is relaxed, the model can duplicate patterns seen in many Paleozoic echinoids with multiple ambulacral plate "columns". The word "columns" is qualified because the plates are not added in columns in a strict sequence one above another, but rather in echelons to form a chevron pattern.

The patterns of ambulacral plates in regular sea urchins are complicated by compounding of individual plates into sets that act in composite as if they were single plates. In effect, the compound ambulacral plate acts in growth in a manner analogous to that of an interambulacral plate, and can be modeled as such. However, from the viewpoint of plate addition, the individual plates of a compound set are inserted individually in accordance with the model described above.

Interambulacral Plate Addition

Interambulacral plates are added to the corona from the external side of the test, usually at the junction of the ocular and the adjacent genital plate (Fig. 5.3a). When a genital plate is missing, the interambulacral plates are added along the adoral edge of the ocular at a point where the genital suture would be expected, rather than at the junction with the adjacent ocular plate (Fig. 5.3b). The ten newly forming interambulacral plates surrounding the apical system of a regular sea urchin have a range of sizes from smallest to largest (Fig. 5.4). These early plates grow at essentially the same rate, therefore the size range implies that smaller plates are younger than larger plates and that the sequence of plate addition is preserved in the sequence of plate sizes. Plates are not added to the columns simultaneously, but there is still a question whether or not plates are added to the columns independently.

A collection of 288 specimens of *S. droebachiensis* (diameters ranging from 12 to 66.5 mm) was examined and the size order of the plates adjacent to the apical system recorded. The relative ages of these plates is closely correlated to the plate size at this early stage of growth. In every case examined, each of the first five smallest interambulacral plates, in order of size, occurred in a separate column, and each of the next five plates also occurred in separate columns. This demonstrates that, although interambulacral plates are not added simultaneously, they are added in cohorts that completely encircle the apical system. This appears to also be the case for ambulacral plates, but has not been confirmed because of the small size of these plates and the fact that they are added from the inside of the test (and are not visible initially).

Considering only the interambulacral plates, a particular sequence [using the Lovén (1874) numbering system] of plate sizes, from smallest to largest, might be:

$$2b \rightarrow 5b \rightarrow 3a \rightarrow 1a \rightarrow 4a \rightarrow 2a \rightarrow 3b \rightarrow 5a \rightarrow 4b \rightarrow 1b$$

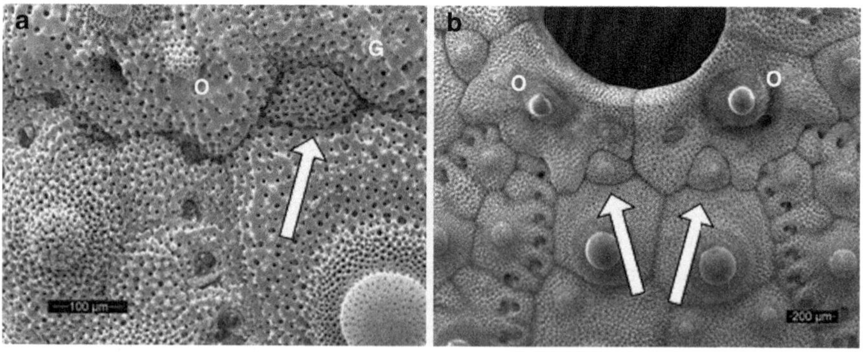

Fig. 5.3 Addition of new interambulacral plates in *Strongylocentrotus droebachiensis*, external view. (**a**) A new plate added at the junction of the ocular (O) and genital (G) plates (*arrow*). (Scale bar 100 µm) (**b**) Genital plate missing. The new interambulacral plates are added at points where junctions with a genital plate would be expected if present (*arrows*) (Scale bar 200 µm)

Fig. 5.4 Apical system of *Strongylocentrotus droebachiensis*. The ten youngest plates (based on size) comprise a cohort that complete encircles the apical system. This particular sequence shows the spiral pattern sometimes seen (starting with the plate in the 5b position and continuing in a counter-clockwise direction 5b→1a→2b→3b→4a→5a→1b→2a→3a→4b)

The actual sequence of plate addition is in the reverse direction:

$$2b \leftarrow 5b \leftarrow 3a \leftarrow 1a \leftarrow 4a \leftarrow 2a \leftarrow 3b \leftarrow 5a \leftarrow 4b \leftarrow 1b$$

And can be represented as a directed graph (Fig. 5.5).

The transition to one position from the previous can be represented by a square matrix of binary values where the value 1 indicates a transition:

	1a	1b	2a	2b	3a	3b	4a	4b	5a	5b
1a	0	0	0	0	0	0	1	0	0	0
1b	0	0	0	1	0	0	0	0	0	0
2a	0	0	0	0	0	1	0	0	0	0
2b	0	0	0	0	0	0	0	0	0	1
3a	1	0	0	0	0	0	0	0	0	0
3b	0	0	0	0	0	0	0	0	1	0
4a	0	0	1	0	0	0	0	0	0	0
4b	0	1	0	0	0	0	0	0	0	0
5a	0	0	0	0	0	0	0	1	0	0
5b	0	0	0	0	1	0	0	0	0	0

Because this represents a directed graph, reading across a row, the transition to a position from the previous is indicated. Reading down a column, the transition from a position to the next position is indicated.

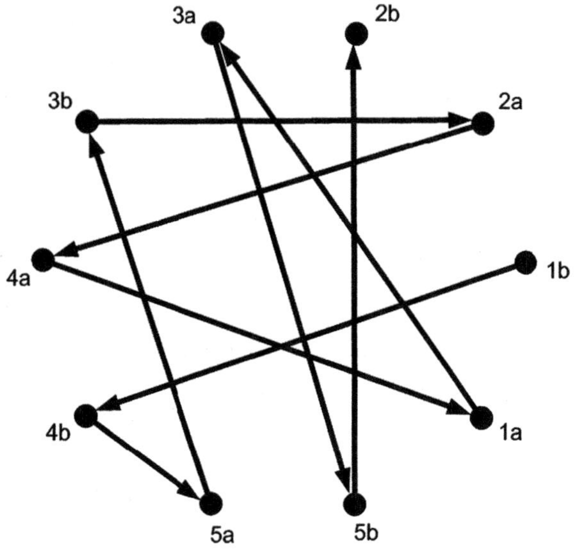

Fig. 5.5 The sequence of plate additions drawn as a directed graph, starting with the 1b position and ending with the 2b position

Each transition table represents the frequency of a transition to one position from another. In an individual case the probabilities are either 0 or 1 for any particular transition, but if the tables for a larger set of specimens are summed and divided by the number of individuals, the resulting transition table approximates the probabilities of the transitions for the sampled population.

The transition table constructed for 288 specimens of *S. droebachiensis* is:

	1a	1b	2a	2b	3a	3b	4a	4b	5a	5b
1a	0.000	0.007	0.188	0.142	0.090	0.128	0.132	0.083	0.135	0.094
1b	0.007	0.000	0.128	0.167	0.122	0.087	0.135	0.139	0.104	0.111
2a	0.108	0.090	0.000	0.010	0.160	0.181	0.101	0.142	0.104	0.104
2b	0.101	0.087	0.007	0.000	0.174	0.135	0.108	0.139	0.115	0.135
3a	0.149	0.118	0.101	0.122	0.000	0.000	0.094	0.174	0.097	0.149
3b	0.111	0.177	0.090	0.111	0.007	0.000	0.184	0.087	0.132	0.101
4a	0.094	0.125	0.125	0.115	0.090	0.149	0.000	0.007	0.184	0.111
4b	0.135	0.108	0.097	0.101	0.149	0.090	0.003	0.000	0.125	0.191
5a	0.128	0.160	0.090	0.142	0.090	0.153	0.146	0.083	0.000	0.003
5b	0.167	0.128	0.174	0.090	0.122	0.076	0.097	0.146	0.000	0.000

The combined column transition probabilities are:

	1	2	3	4	5
1	0.007	0.312	0.214	0.245	0.222
2	0.193	0.009	0.325	0.245	0.229
3	0.278	0.212	0.003	0.269	0.240
4	0.231	0.219	0.240	0.005	0.306
5	0.292	0.248	0.220	0.236	0.002

If the addition of plates acts independently the position relative to the apical system is unimportant and plates in any column have an equal probability of insertion and the expected transition frequencies are:

	1	2	3	4	5
1	0.000	0.250	0.250	0.250	0.250
2	0.250	0.000	0.250	0.250	0.250
3	0.250	0.250	0.000	0.250	0.250
4	0.250	0.250	0.250	0.000	0.250
5	0.250	0.250	0.250	0.250	0.000

The positive residuals between the observed and expected values are not large, but indicate a subtle preference for a 1→2→3→4→5 sequence (from smallest to largest).

	1	2	3	4	5
5	0.042	-	-	-	-
1	-	0.062	-	-	-
2	-	-	0.075	-	-
3	0.028	-	-	0.019	-
4	-	-	-	-	0.056

A tendency towards a particular sequence would reveal itself as a recurrent series. For example, if a 1→2→3→4→5 sequence were present, recurrent series would be 1–2–3–4–5, 2–3–4–5–1, 3–4–5–1–2, etc. The multiple independent series can be consolidated by rotating the individual sequences so that each begins with 1. Using this consolidation there are 24 different possible sequences. The frequencies with which these sequences were observed in the sample are:

Sequence	Frequency	Expected	Residual
1-2-3-4-5	0.084	0.042	0.042
1-2-3-5-4	0.035	0.042	-0.007
1-2-4-3-5	0.035	0.042	-0.007
1-2-4-5-3	0.045	0.042	0.003
1-2-5-3-4	0.042	0.042	0.000
1-2-5-4-3	0.042	0.042	0.000
1-3-2-4-5	0.049	0.042	0.007
1-3-2-5-4	0.035	0.042	-0.007
1-3-4-2-5	0.031	0.042	-0.011
1-3-4-5-2	0.042	0.042	0.000
1-3-5-2-4	0.045	0.042	0.003
1-3-5-4-2	0.042	0.042	0.000
1-4-2-3-5	0.035	0.042	-0.007
1-4-2-5-3	0.038	0.042	-0.004
1-4-3-2-5	0.031	0.042	-0.011
1-4-3-5-2	0.017	0.042	-0.025
1-4-5-2-3	0.070	0.042	0.028
1-4-5-3-2	0.049	0.042	0.007
1-5-2-3-4	0.049	0.042	0.007
1-5-2-4-3	0.031	0.042	-0.011
1-5-3-2-4	0.028	0.042	-0.014
1-5-3-4-2	0.035	0.042	-0.007
1-5-4-2-3	0.042	0.042	0.000
1-5-4-3-2	0.045	0.042	0.003

In general, the frequencies are not much different from the expected values, but the same sequence noted earlier (1–2–3–4–5) occurs at twice the expected frequency.

This is the basis for the spiral pattern alluded to by Agassiz (1834), with an example of this pattern shown in Fig. 5.4. However, this particular sequence still only accounts for 8.4% of the sequences expressed by this species of urchin, and the overall pattern supports the hypothesis of independence of the interambulacral columns.

The independence of the interambulacral columns is evidence that the process underlying the addition of new plates arises from properties associated directly with the columns and their geometry. This criterion is met by the threshold model described by Zachos (2009), in which morphogens generated by the growing coronal plates inhibit the production of new plates when concentrations at an insertion point exceed a threshold level as described by the equation:

$$C_i = \left(1 - \frac{A_t}{A_{\max}}\right) k/d^2$$

where C_i is the concentration of the morphogen produced by plate i, A_t and $A_{\max}$ are the area of plate i at time t and at maximum size, respectively, d is the Great Circle distance between the center of plate i and the point of plate insertion, and k is a constant representing maximum concentration. The concentration of morphogen at any given insertion point is the sum of concentrations from all surrounding coronal plates, and when this concentration falls below a threshold value a new plate is added at that point.

Plate Growth

The term accretion has been applied to the growth of individual plates of echinoids (Smith 2005), but in general usage the term implies that growth proceeds by addition of discrete elements or particles to a plate, which is misleading. Rather than accreting skeletal material, the stereom of a plate expands in a dendritic pattern, analogous in many ways to the growth of filamentous colonies of fungi (Edelstein 1982). Information regarding the growth of individual plates of modern echinoids is preserved by growth lines in the plates (Deutler 1926). It has been shown (Zachos 2009) that the growth can be modeled by the linear form of the Bertalanffy equation (Bertalanffy 1938) which can be expressed as:

$$l_t = L - (L - l_0)e^{-kt}$$

where l_t is the length of the plate perimeter at time t, l_0 the initial length, L the maximum length attained, and k the growth rate.

Indistinct growth banding has been reported for the Paleozoic genus *Archaeocidaris* (Smith 2005). Rarely, distinct growth banding can be seen in the interambulacral plates of the palaechinid *Lovenechinus missouriensis* (Fig. 5.6a), and is strong evidence that plates in the family Palaechinidae grew laterally in the same manner as seen in modern echinoids even though these Paleozoic species are

Fig. 5.6 Preservation of growth rings in Paleozoic echinoids. (**a**) *Lovenechinus missouriensis* (MCZ 101944). (Scale bar 5 mm) (**b**) *Lepidocentrus mulleri* (MCZ 101877) (Scale bar 1 mm)

characterized by very thick, robust plates. A number of disarticulated plates of *Lepidocentrus mulleri* are characterized by light and dark color bands (Fig. 5.6b) that appear to represent growth bands. If so, they indicate that rather than demonstrating approximate isotropic growth from the center (as seen in modern echinoids and apparently in the Palaechinidae and Archaeocidaridae), these plates grew asymmetrically from one corner in the direction of the strongly overlapping imbrications of these plates.

Models of Plate Addition and Growth

Crown Group Regular Echinoids

A new methodology for modeling sea urchin skeletal growth combines plate addition and growth in a spherical (3D) reference frame (Zachos 2009). Plates are represented by growth centers that meet the criteria for Delaunay triangulation, and the plates themselves can be represented by the dual Voronoi polygons for simplicity.

Different models are used for the location of ambulacral and interambulacral plate addition loci. The underlying water vascular system is modeled by locating the radial canal along a line extending from the center of the mouth to the junction of two ambulacral plates and an ocular. Insertion loci for ambulacral plates are located at a variable distance perpendicular to the radial canal (representing the length of the lateral canals), and at a variable distance from the center of the ocular plate (representing the terminal end of the radial canal). Insertion loci for interambulacral plates are located along a line extending from the genital – ocular – interambulacral plate junction to the center of the interambulacral plate. The plates are assumed to be infinitely thin and there is no distinction between plate addition from the inside or outside of the test.

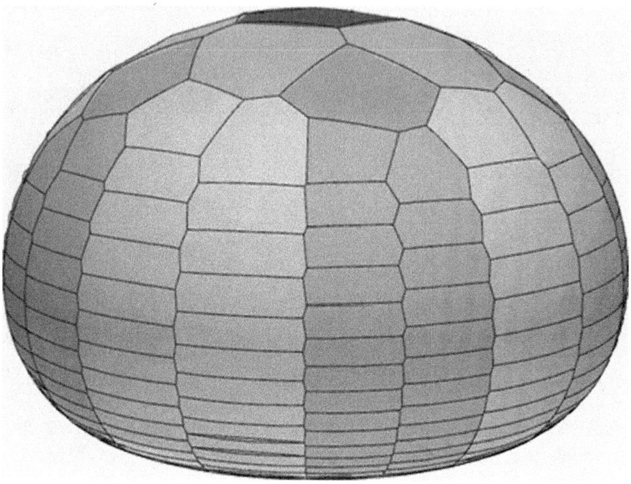

Fig. 5.7 Example model of a generic modern regular echinoid

The Bertalanffy model is used for plate growth. The best results are obtained when the ambulacral and interambulacral plates have the same growth rate, but different maximum perimeter values and plate addition rates. The resulting models are sensitive to initial conditions, but in most cases quickly converge on a relatively simple pattern of alternating plate addition and growth. Although the calculations are made over a spherical frame of reference, the resulting models can be deformed (Fig. 5.7) to display a typical sea urchin shape (Ellers 1993).

Stem Group Paleozoic Echinoids

Among several principles used to understand the morphology of echinoids, one of the more pervasive is that all interambulacral plates originate in direct contact with the ocular plates (Jackson 1912). This principle is called the Ocular Plate Rule or OPR (David et al. 1995; Mooi and David 1993, 1997; Mooi et al. 1994). This principle has been validated for all crown group echinoids. In nearly all cases the interambulacral plates are introduced at the junction of the genital and ocular plates, exceptions occurring in the Echinothurioida (in which the genital and ocular plates may not be in contact) and in sporadic teratologies of the apical system. The key assumption of the OPR is that there are only two loci for new plate addition in each interradius. Because all modern echinoids have only two interambulacral plate columns per interradius, it is no surprise that the OPR is observed. It has either been explicitly stated (Jackson 1912) or implied (Mooi and David 1993) that the OPR is valid not only for modern echinoids, but also for the multicolumn Paleozoic echinoids. This assertion, however, has never been adequately tested for the Paleozoic forms.

A number of attempts have been made to classify the genera of echinoids known from the Paleozoic (Jackson 1912; Kier 1966; Smith 1984), but at best they are grouped as stem group echinoids with uncertain relationships (Smith 2010). However, for the purposes of geometric modeling of the skeleton, it is convenient to divide the majority of the genera into three major groups (Kier 1965): Echinocystitoida, Palaechinoida (restricted to the family Palaechinidae), and Cidaroida (restricted to the family Archaeocidaridae). This general division excludes the problematic Bothriocidaridae, other somewhat cryptic genera, and the Miocidaridae (which are modeled as modern forms). The species in the Palaechinidae are all characterized by relatively thick, robust, non-imbricated plates. Although the species have variable numbers of both ambulacral and interambulacral columns, the lack of imbrication makes them, from a geometric modeling standpoint, the simplest to study. The Echinocystitoida include genera with imbricate plating, as well as forms (e.g., *Lepidesthes* and *Meekechinus*) with greatly expanded ambulacral zones. The Archaeocidaridae are also characterized by imbrication, but with a significantly lesser degree of overlap and strictly biserial ambulacral plating.

Echinocystitoida

The forms with strongly imbricate interambulacral plates present the most difficult problems for modeling and no satisfactory 3D model has yet been developed. The plates, at least in some cases, do not grow isotropically from the center but rather in an asymmetric fashion. More importantly, because of significant plate overlap the plate centers do not meet the criteria for a Delaunay triangulation. However, some of the forms with imbricate ambulacral plates can be represented by simulating the growth of the lateral canals of the water vascular system rather than the growth of the plates directly. The species *Lepidesthes colleti* (Fig. 5.8a) is an excellent example for demonstrating how the model operates (Fig. 5.8b).

Palaechinoida

This division, composed of the genera in the family Palaechinidae, is characterized by forms with thick, robust plates and a semi-rigid, apparently nearly spherical body form. The plates appear to have reached near maximum thickness as soon as they are added and further growth was by accretion on the interplate sutures (Fig. 5.6a). The ambulacra were biserial in some genera (*Palaechinus*, *Maccoya*) or, if multiserial, constant and relatively simple in structure. The sizing pattern of plating around the apical system shows the apparent addition of interambulacral plates along the adoral edge of the genital plates without contact with the oculars (Fig. 5.9). This suggests that the same model as used for growth of regular echinoids can be used for palaechinids, with modifications to allow additional interambulacral insertion points marginal to each genital plate. This is an explicit violation of the Ocular Plate Rule, but even simple models explain some of the characteristic

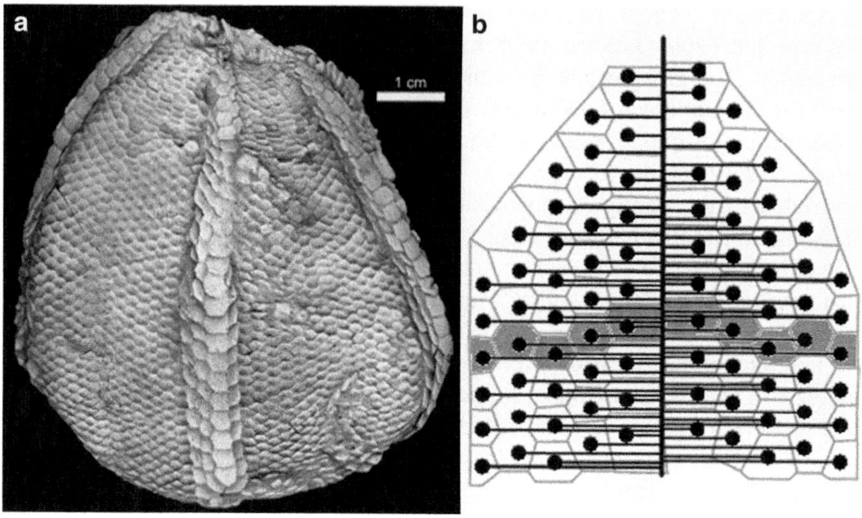

Fig. 5.8 Modeling growth of multicolumned ambulacral plates. (**a**) *Lepidesthes colleti* (MCZ 101991), showing multicolumn pattern in ambulacra. (Scale bar 1 cm) (**b**) Representation of water vascular system with single radial canal and left-right alternating lateral canals. One cohort set of plates is shaded

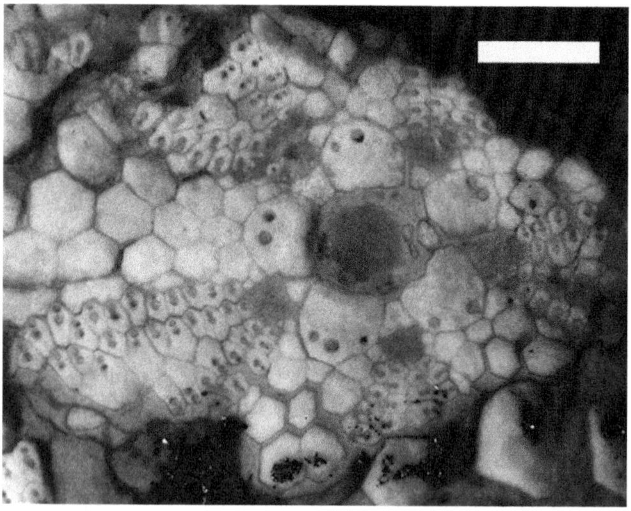

Fig. 5.9 Apical region of *Melonechinus mulitporus* (USNM S3851). Some of the young (small) interambulacral plates were added from points adjacent to the genital plates only (Scale bar 3 mm)

plate patterns seen in the interambulacral columns of these echinoids. The ultimate number of columns in an interambulacrum is not absolutely dependent upon the number of insertion points, but is sensitive to their placement and separation.

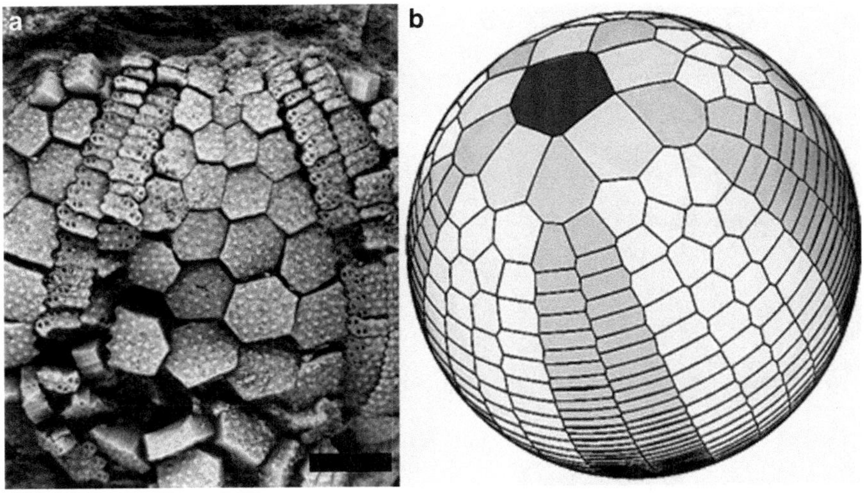

Fig. 5.10 Comparison of plate patterns in Paleozoic echinoids and models. (**a**) *Maccoya burlingtonensis* (USNM S3840), external view. (Scale bar 5 mm) The plates heading the intercalated columns are often pentagonal in shape. (**b**) Model using four interambulacral insertion points

This could be a cause of reported variability in the number and pattern of interambulacral columns (Klem 1904). Jackson (1896, 1912) noted that pentagonal plates were at either end of intercalated columns that appeared to begin within a plate mosaic away from the apical system (Fig. 5.10a). This feature is duplicated in the models, and is an artifact of the underlying Delaunay triangulation of the plate centers and further supports this model (Fig. 5.10b). The diamond-shaped plates which often occur "orphaned" in the plate mosaic of palaechinids (Fig. 5.11a) are also duplicated in the model (Fig. 5.11b). They arise naturally when growth of the surrounding plates causes a separation of the plate cohorts. The pattern of migration of the plate cohorts from the apical pole towards the mouth is explicitly recorded in the models, and demonstrates that in the actual fossils the relative ages of the plates can be interpreted in exactly the same fashion. In effect, it is geometrically impossible within the assumptions of the model to have the columns all originate in strict accordance with the Ocular Plate Rule.

Cidaroida

This group, composed of the genera in the family Archaeocidaridae, most closely resembles modern echinoids. Although the plates are imbricate and the skeleton was apparently flexible to some degree, it did not differ geometrically from that of modern cidaroids except in the number of interambulacral columns. It has not been possible to locate any specimens of any of the species in this family with a well-preserved apical region, and therefore the details of the apical system and the patterns of newly added interambulcral plates remain unknown. Because the plates

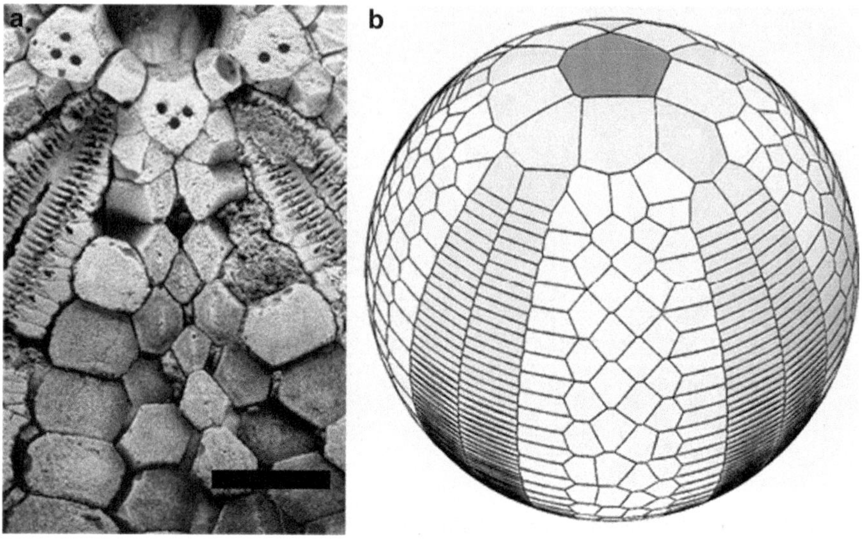

Fig. 5.11 Comparison of plate patterns in Paleozoic echinoids and models. (**a**) *Lovenechinus missouriensis* (MCZ 101944), internal view, latex cast. (Scale bar 5 mm) Orphaned plates are typically diamond-shaped where surrounded by other plates from older cohorts until they grow to form a continuous column. (**b**) Model using six interambulacral plate insertion points

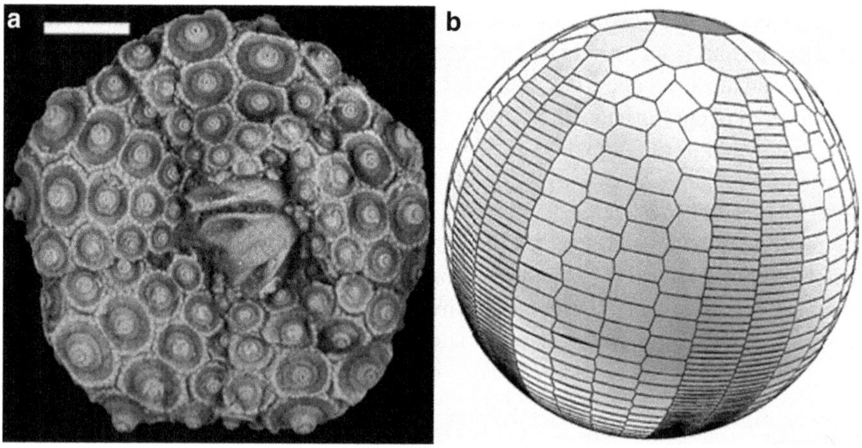

Fig. 5.12 Comparison of plate patterns in Paleozoic echinoids and models. (**a**) *Archaeocidaris legrandensis* (USNM S3828), oral side. (Scale bar 5 mm) (**b**) Model using four interambulacral insertion points

are imbricate, it is possible that the origin of inner columns of the interradii could have occurred within the constraints of the Ocular Plate Rule. However, it is also possible to realistically model the columns with multiple insertion points (Fig. 5.12a, b).

Conclusions

The echinoid skeleton offers a unique opportunity for computational modeling. The processes of growth are well understood, at least conceptually. The geometry of the skeleton is simple in design, but complex in construction. A relatively simple computational geometric model can simulate the general patterns of growth in modern sea urchins. A model of echinoid growth using a Bertalanffy growth model for plate growth and an inhibition-activation model for plate addition successfully reproduces many patterns characteristic of modern echinoid skeletons. Modern echinoids invariably have 20 columns of plates, arranged in biserial columns in each of five radial and five interradial areas. Paleozoic echinoids appear odd because of their range of morphological disparity characterized by the number of plate columns (15 to over 150) but the same model for growth of individual plates can be applied to both modern and Paleozoic echinoids. The assumption of isotropic plate growth is valid except for some types of imbricate plating. The same inhibition-activation model is valid for Paleozoic forms, although requiring additional plate insertion loci to reproduce multiple plate "columns" in the interradii. Patterns of plates in the radii can be reproduced using a model of the underlying water vascular "tree" to guide plate accommodation caused by expansion of the radial zones during growth. Modeling results reproduce many otherwise unexplained features observed in Paleozoic skeletons, including orphaned plates, pentagonal plates, and intercalated columns.

Acknowledgements We would like to thank collections managers Jann Thompson (National Museum of Natural History), Jessica Cundiff (Harvard University Museum of Comparative Zoology), Ann Molineux (Texas Natural Science Center), and Paul Mayer (Field Museum) for access to their collections of Paleozoic echinoids. Douglas Erwin and David Pawson sponsored the senior author during a postdoctoral appointment at the National Museum of Natural History for research on this project. Thanks are due Scott Whitaker, National Museum of Natural History, for all of his help with ESEM and optical microscopy. We are grateful to Hank Stence, R.J. Peacock Hatchery, Lubec, Maine for young *S. droebachiensis* and many pleasant hours watching his sea urchins grow. Richard Reyment helped improve the manuscript with his comments. This research was supported by the Smithsonian Institution Office of Fellowships and the Jackson School of Geosciences, the University of Texas at Austin.

References

Agassiz, L., 1834. Observations on the growth and on the bilateral symmetry of Echinodermata. London and Edinburgh Philosophical Magazine and Journal of Science 5, 369-372.

Bertalanffy, L. von, 1938. A quantitative theory of organic growth (inquiries on growth laws. II). Human Biology 10, 181-213.

Bookstein, F.L., 1996. Biometrics, biomathematics and the morphometric synthesis. Bulletin of Mathematical Biology 58, 313-365.

Cavey, M.J., and Märkel, K., 1994. Echinoidea, in: Harrison, F.W. and Chia, F.S., (Eds.), Microscopic Anatomy of Invertebrates: Echinodermata, Vol. 14. Wiley-Liss, Inc., New York, pp. 345-400.

David, B., Mooi, R., and Telford, M., 1995. The ontogenetic basis of Loven's Rule clarifies homologies of the echinoid peristome, in: Emson, S.C., (Ed.), Echinoderm Research 1995, A.A. Balkema, Rotterdam, pp. 155-164.
Deutler, F., 1926. Über das Wachstum des Seeigelskeletts. Zoologische Jahrbücher Abteilung für Anatomie und Ontogenie der Tiere 48, 119-200.
Edelstein, L., 1982. The propagation of fungal colonies: a model for tissue growth. Journal of Theoretical Biology 98, 679-701.
Ellers, O., 1993. A mechanical model of growth in regular sea urchins: predictions of shape and a developmental morphospace. Proceedings of the Royal Society of London B 254, 123-129.
Ellers, O., Johnson, A.S., and Moberg, P.E., 1998. Structural strengthening of urchin skeletons by collagenous sutural ligaments. Biological Bulletin 195, 136-144.
Gordon, I., 1926. The development of the calcareous test of *Echinus miliaris*. Royal Society of London, Philosophical Transactions, ser. B 214, 259-312.
Hall, B.K., 1992. Evolutionary Developmental Biology. Chapman & Hall, London, 275p.
Jackson, R.T., 1896. Studies of Palæechinoidea. Geological Society of America Bulletin 7, 171-254.
Jackson, R.T., 1912. Phylogeny of the Echini, with a revision of the Paleozoic species. Boston Society of Natural History, Memoir 7, 491p.
Kier, P.M., 1965. Evolutionary trends in Paleozoic echinoids. Journal of Paleontology 39, 436-465.
Kier, P.M., 1966. Noncidaroid Paleozoic Echinoids, in: Moore, R.C., (Ed.), Treatise On Invertebrate Paleontology. Part U. Echinodermata 3, Vol. 1. Geological Society of America and University of Kansas Press, New York and Lawrence, pp. U298-U312.
Kitano, H., 2002. Computational systems biology. Nature 420, 206-210.
Klem, M.J., 1904. A revison of the Paleozoic Palaeechinoidea, with a synopsis of all known species. Transactions of the Academy of Science of St. Louis 14, 1-98.
Konarzewski, M., Kooijman, S.A.L.M., and Ricklefs, R.E., 1998, Models for growth, in: Starck, J.M. and Ricklefs, R.E., (Eds.), Avian Growth and Development, Oxford University Press, Oxford, pp. 340-365.
Kumar, S., and Bentley, P.J. (Eds.), 2002. On Growth, Form and Computers. Elsevier, London.
Lovén, S., 1874. Études sur les Échinoïdées. Konglingar Svenska Vetenskaps-Akademiens Handlingar 11, 1-91.
Märkel, K., 1981. Experimental morphology of coronar growth in regular echinoids. Zoomorphology 97, 31-52.
Märkel, K., Röser, U., and Stauber, M., 1989. On the ultrastructure and the supposed function of the mineralizing matrix coat of sea urchins (Echinodermata, Echinoida). Zoomorphology 109, 79-87.
McGhee, G.R., Jr, 1999. Theoretical Morphology. Columbia University Press, New York, 316p.
Mooi, R., and David, B., 1993. Novel skeletal topologies are related to birth in Antarctic sea urchins. Comptes rendus de l'Academie des sciences. Serie III, Sciences de la vie 316, 341–345.
Mooi, R., and David, B., 1997. Skeletal homologies of echinoderms, in: Waters, J.A. and Maples, C.G., (Eds.), Geobiology of Echinoderms, Paleontological Society Papers, Vol. 3, pp. 305-335.
Mooi, R., David, B., and Marchand, D., 1994. Echinoderm skeletal homologies: Classical morphology meets modern phylogenetics, in: David, B., et al., (Eds.), Echinoderms through Time, A.A. Balkema, Rotterdam, pp. 87-95.
Reif, W.-E., and Weishampel, D.B., 1991. Theoretical morphology, an annotated bibliography 1960-1990. Courier Forschungsinstitut Senckenberg 142, 140p.
Smith, A., 1980. Stereom microstructure of the echinoid test. Special Papers in Palaeontology, Vol. 25, The Palaeontological Association, London, 81p.
Smith, A., 1984. Echinoid Palaeobiology. George Allen & Unwin, London, 190p.
Smith, A., 2005. Growth and form in echinoids: the evolutionary interplay of plate accretion and plate addition, in: Briggs, D.E.G., (Ed.), Evolving Form and Function: Fossils and Development, Yale Peabody Museum Publications, New Haven, pp. 181-195.

Smith, A. (Ed.), 2010. The Echinoid Directory. World Wide Web electronic publication. http://www.nhm.ac.uk/research-curation/projects/echinoid-directory/index [accessed March 1, 2010].

Wilkins, A.S., 2002. The Evolution of Developmental Pathways. Sinauer Associates, Inc., Sunderland, MA, 603p.

Zachos, L.G., 2009. A new computational growth model for sea urchin skeletons. Journal of Theoretical Biology 259, 646-657.

Chapter 6
Morphometric Analysis of Polyphenism in Lower Cretaceous Ammonite Genus *Knemiceras*

Richard A. Reyment

Introduction

Many ammonite taxa, commonly considered to be genetically homogeneous, can display a surprising degree of morphological diversity. One such genus is *Knemiceras* of Albian age. In addition to variability in shell-shape due to unspecified ecological factors there is the question of variability arising from sexual dimorphism in the sense of Makowski (1963) and Callomon (1963) which in essence relates sex-differences to size and ornament. The terms "microconchs" (tentatively taken to be males) and "macroconchs", taken to be females are applied in an ad hoc manner. These shells are identified with respect to maturity by looking for the crowding of the final septa. The accepted opinion is that when an animal was approaching maturity, adjustments were made to the final buoyancy which involved the secretion of one or more chambers of lesser volume. An analogy can be drawn with the *Nautilus* where usually one final chamber is "smaller" than the immediately preceding ones; it is however necessary to point out that ammonites and nautiloids both secrete chambered shells, but they are not very "close relatives". It is therefore necessary to exercise caution respecting attempts at extrapolating information obtained for one category of cephalopods to the other. This has been demonstrated by Mutvei and Dunca (2007, pp. 240–241).

A problem in identifying adult shells in ammonites is that sutural crowding can only be observed on shells, the outer calcareous layer of shell-material of which has been lost during diagenesis. There are mineralogical reasons why shell material remains, either as original aragonite or transformed to calcite, a situation that complicates the identification of adult conchs. One must then seek alternative aids to achieve this end, one of which is to look for umbilical egression [the diameter of the umbilicus tends to widen on the final whorl of an shell approaching adulthood Urlichs (2006)], another is to attempt to assess the onset of pronounced changes in adult ornament. An indicator that is also of diagnostic value is the size

R.A. Reyment
Naturhistoriska Riksmuset, 10405, Stockholm, Sweden
e-mail: richard.reyment@nrm.se

A.M.T. Elewa (ed.), *Computational Paleontology*,
DOI 10.1007/978-3-642-16271-8_6, © Springer-Verlag Berlin Heidelberg 2011

of a specimen. In many cases, microconchs are essentially smaller than macroconchs. However this is not a hard and fast rule and the size of some microconchs of *Knemiceras persicum* Collignon differ but slightly from macroconchs. Thus size differentiation on its own does not strike me as being uniquely adequate for the purposes of taxonomic classification and there seems to be a need for a more exact modus operandi such is available from the corpus of multivariate morphometric analysis (*sensu* Blackith 1965).

Davis et al. (1996, p. 521) ended their review of mature modifications and sexual dimorphism in ammonoids by putting several propositions for general consideration, for workers at large, within the framework of the eventual influence of environmental factors with reference to the following standpoints: (1) Are dimorphic species more eurytopic than monomorphic ones? (2) Are dimorphic species more abundant (than those not displaying identifiable dimorphic characteristics)? (3) Did dimorphic species produce a greater number of individuals? Davis et al. (1996) noted also that almost all Cretaceous species interpreted as being dimorphic are shallow to mid-shelf forms. Did then ecological conditions play a significant part in the expression of shell-forms? (In a written reply to a question from me in 2009, Professor Urlichs kindly confirmed much the same palaeoecological situation seem to apply for ceratitids.) The problems identified by Davis et al. (1996), although highly relevant, do not cover the entire scope of interpreting infraspecific variability in the conchs of certain ammonite species. The present study makes use of multivariate statistical procedures applied to a well-studied ammonite species in an attempt att elucidating some of the problems formulated by Davis et al. (1996). Reyment (2003, 2004) described and analysed statistically pronounced variability in shell-shape of Turonian (Upper Cretaceous) ammonites from the southern Sahara. In these studies it was deduced that there is a wide range of shape differentiation within a single species. Most of these variations were originally considered at that time to be different species (cf. Barber 1957), but it now seems that all of these forms can be housed under just a few taxa, even including remarkable sports, such as "*Eotissotia*", which has an analogue in *K. persicum*. The same situation (although more extremely manifested) occurs even to the extent of analogous sports which on first encounter seem to be quite separate. In the spirit of the theses of Davis et al. (1996) our interest lies with exemplifying possible interpretations of polymorphism in *K. persicum* Collignon. Reyment and Kennedy (1991) examined the same material as consulted here, and arrived at the conclusion that it could possibly represent a case of phenotypic plasticity such as could be expected in a labile (epicontinental) environment (Via and Lande 1985). Swan and Saunders (1987) applied raw principal component analysis to Palaeozoic ammonoids in an attempt to establish a statistical relationship between function and shape. As pointed out by Reyment and Kennedy (1991, p. 412), the analysis suffers from serious weaknesses, including indeterminancy arising out of the use of compositional data in the wrong multidimensional setting, that is, Cartesian space instead of Simplex space and, furthermore, redundancy in some of the measures made (cf. Aitchison 1986). These faults conspire so as to invalidate any conclusions ventured on the basis of the "statistical results".

Morphometric Methods

A useful tactic for sifting the information carried by a set of taxonomically selected variables is that of cross-validation. Cross-validation has attained prominence owing to its growing importance in quantitative analytical chemistry (Wold 1978). The approach used in the following is that due to Krzanowski (1982, 1987a, b), exemplified for palaeontological applications by Reyment (1991) and Reyment and Savazzi (1999). The correlations between variables were used for computating the latent roots and vectors for statistical reasons as outlined in Krzanowski (1982), thus reducing the problem to working with hyperspheres and not hyperellipsoids. So far, the methodology of cross-validation has been of a somewhat informal nature in that exact tests have not yet been forthcoming for establishing objective significance criteria (Krzanowski 1987a, b).

Topics suitable for testing by cross-validation are:

1. Can any of the variables be excluded on the grounds of redundancy? In palaeontology this can be a two-edged sword since statistical relevance does not always accord with paleobiological expedience and logic.
2. How many principal components contribute essential information?
3. Do any of the observations deviate in some multivariate analytical perspective from the main body of the data and which is not evidenced in a bivariate scatter-plot?

Cross-validation is essentially an exploratory technique (currently referred to as data-mining in the literature) for finding informative patterns and "scans" a data-matrix for redundancy. The steps in performing the calculations are:

1. Compute the principal components of the covariance matrix **S**, or the correlation matrix **R**, of the $(n \times p)$ data-matrix **X**. These are yielded as the solution of the equation $\mathbf{S} = \mathbf{VLV*}$ such that $\mathbf{V*V} = 1$. Alternatively, the singular value decomposition of **X** can be applied, to wit, $\mathbf{X} = \mathbf{UDV*}$, **D** diagonal.
2. Compute the scores of the principal components $\mathbf{Z} = \mathbf{XV*}$. (N.B. the asterisk is here used to denote a transposed matrix).
3. Determine Krzanowski's criterion W_m, which is obtained from the average squared discrepancy between the actual values and "predicted" values of the data-matrix.
4. Divide the data-matrix into several groups. Delete then each such group separately from the data-matrix and compute the values of the predictor from the remainder [cf. application by Reyment (2004, p. 634)].
5. Informative values: The comparison of two m-dimensional configurations of the same n points may be done by means of "Procrustean Analysis" (Gower 1975) by which the sum of distances between corresponding points of the two configurations is found after matching under translation, rotation and reflection. A large sum of squares resulting from the suppression of variable x_i indicates a discrepancy between the two configurations, thus suggesting that this variable is probably contributing essential information to the analysis. The alternative

result would indicate that there is a close match between the two configurations which means that the removal of the variable in question from the analysis does not bring about an undue loss of efficiency. On purely statistical grounds a variate may seem eligible for exclusion but not on palaeontological. A case of this quandary in the present study concerns the development of the umbilicus as echoed in the final stages of growth of the umbilicus.

The Influence of Individual Specimens

We have now arrived at the main point of interest in the palaeontological application of cross-validation, notably, the influence of each individual in the sample. This analysis is in effect a means of identifying conchs that deviate from the norm as represented by the entire set of juveniles, macroconchs and microconchs. There are two types of divergent observations that are of taxonomic significance. An *atypical* observation is one that deviates strongly from the rest of the sample and hence may exert an unwarranted influence on the analysis. An *influential* individual is one that causes a pronounced change in the computations when it is excluded from the analysis but the suite of measurements on it show no obvious divergencies from the data-set as a whole.

Principal Coordinates

A second diagnostic procedure is that of principal coordinate analysis (Gower 1966) which is a Q-mode method for exploiting statistical distances (taxonomic in the present case) with the end in view of producing a graphical representation of the data. The name given the procedure by Gower is a calque on the term "principal components" . Principal coordinates, using Gower's similarity matrix, permits the union of quantitative, qualitative and dichotomous variables in the same analysis. Attempting to reify the distance based vectors of principal coordinates is not a procedure to be encouraged, being based on a false analogy with what is often done in principal component analysis.

The Data

The data consist of ten "distance variables", that is measurements on dimensions of the conch. Six of these measure dimensions of the apertural face of the conch and four lateral dimensions, as indicated in Fig. 6.1. More precisely, these are

The Apertural Set: (1) maximum diameter of conch, (2) ventral breadth of last whorl, (3) maximum breadth of the last whorl, (4) minimum ventral breadth of the

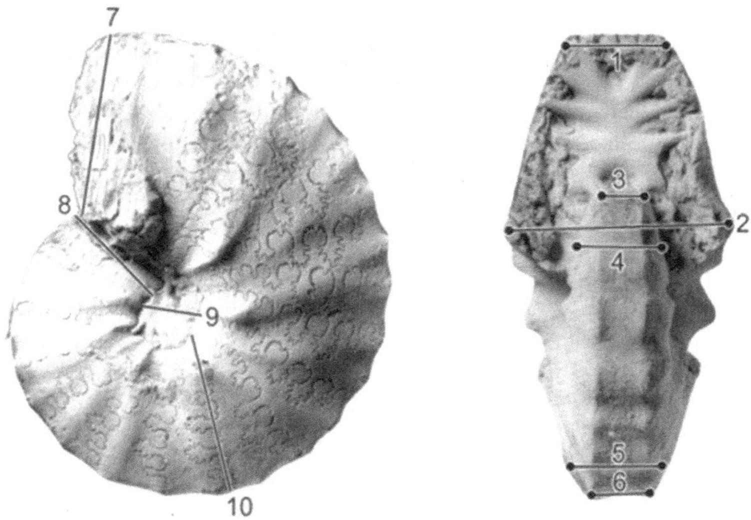

Fig. 6.1 The ten distance characteristics measured on the conchs

second last whorl, (5) maximum ventral breadth of the second last whorl, (6) minimum ventral breadth of the last whorl.

The Lateral Set: (7) Maximum height of the last whorl, (8) distance of the last whorl recorded at right angles to the aperture to the umbilical boundary, (9) the breadth of the umbilicus measured from the intersection of (8), (10) the width of the last whorl on the diametrically opposite side to the intersection of (7) with (8).

The number of measurable conchs is not great granted that the entire collection contains more than 100 specimens. However, as Urlichs (2006) stresses, interpretations of the scope of dimorphism in ammonites cannot be relied upon if defective specimens are used. As a preliminary to the quantitative analyses, it is instructive to review Urlichs (2009) in his presentation of the salient features to be observed in establishing dimorphism in ammonoids, made a useful point with respect to the egression (i.e., relaxation of the angle of coiling – "uncoiling") of the umbilicus from the phragmocone to the end of the living chamber. This is a morphological development usually relatable to the onset of maturity and in the present analysis proven itself to be useful. Makowski's presumed sexual dimorphism in Jurassic ammonites to be expressed by the following criteria (1) The two categories should have identical ontogenetic histories. (2) There should be a lack of intermediary conch shapes between the two categories. (3) Both morphs should occur in the same strata.

Additionally, the illustrations accompanying Makowski's monograph show that the two size categories usually have different ornamental features, in addition to lappets on the microconch. Reyment (1988, p. 110) observed that Makowski (1963) original presentation of dimorphic characteristics contains examples where microconchs and macroconchs, considered to be of the same species, deviate from the third requirement in that they occur separated in time. Ornamental dimorphism is known to occur (Urlichs 2009, p. 200). In ceratites, the ornament may change character at the termination of the phragmocone of macroconchs, whereas microconchs remain unchanged. The Turonian (Cretaceous) genus *Benueites* is an example of well presented ornamental dimorphism (Reyment 1971). Two differing morphologies are illustrated in Fig. 6.2, the one compressed, the other inflated and trapezoidal in whorl section.

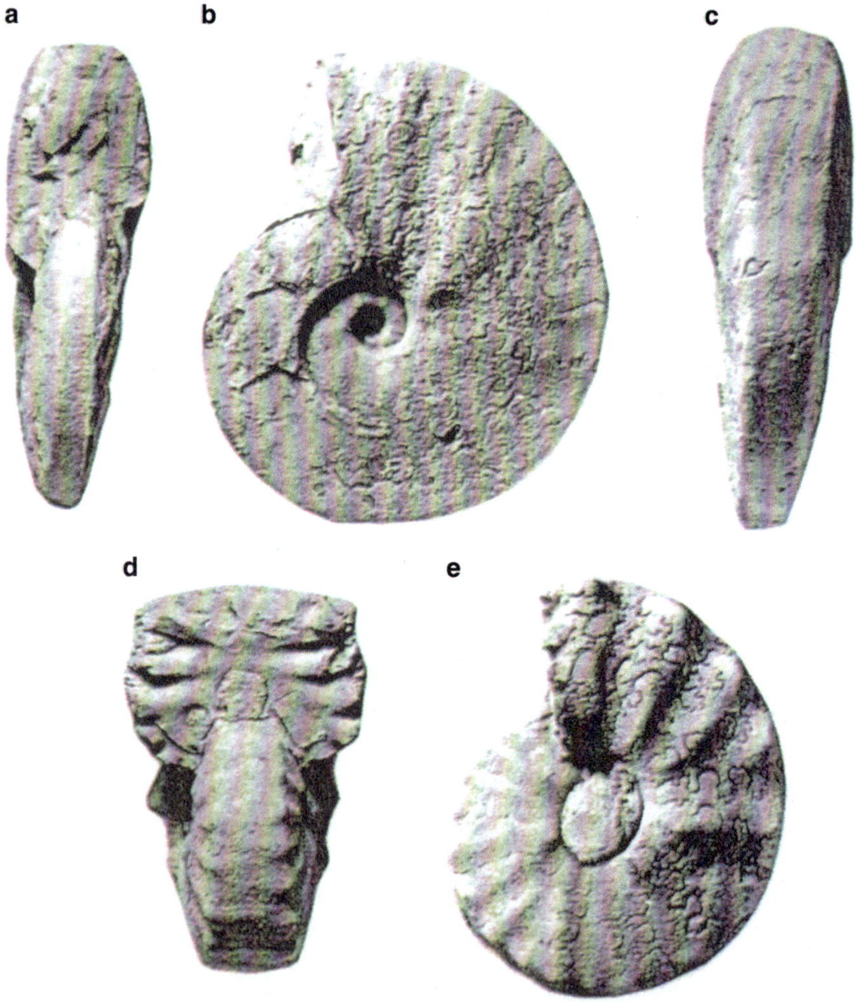

Fig. 6.2 Two specimens of *K. persicum* illustrating extremes in shell shape. (**a–c**) a microconch shell, OUM KY49, diameter 77.2 mm. View (**c**) shows the egression of the umbilicus outlined by the impression of the last part of the last whorl. (**d–e**) is a juvenile shell with a very depressed whorl section, OUM KY7, diameter 39 mm. OUM = Oxford University Museum

The use of septal crowding as a means of ascertaining the maturity of a conch has a natural limitation in that sutures are only exposed on moulds, that is, specimens lacking the external shell material (either as original aragonite or transformed to calcite). Thus, the material studied by Reyment (1971) of *Benueites*, is excellently preserved and not one suture has been observable on any of the specimens owing to the preservation of the outer shell (as calcite). Clearly, in such cases it is desirable to have some supplementary means of evaluating the status of a conch. This endeavour forms one of the principal aims of the present contribution.

Mutvei and Dunca (2007, p. 252) concluded that, despite claims arising from untested dogma, ammonites were not jet-powered swimmers as, for example, is indicated by the fact that the volume of water in the ventral mantle cavity was insufficient for jet-propulsion, among other morphological features. These authors also emphasized that considerations of buoyancy changes in ammonites, extrapolated from living *Nautilus* are based on a false biofunctional analogy. Reyment (1988) used mathematical energy conservation models to show that most shell-types occurring among ammonites were not compatible with a foraging technique requiring rapid deplacement due to energy requirements.

Statistical Analysis

The number of components worth keeping can be gauged by computing PRESS, the Prediction Sum of Squares, a method proposed by Krzanowski (1987b, p. 578), and exemplified in a geological connexion in Reyment and Jöreskog (1993, p. 116) and Reyment and Savazzi (1999) to which the reader is referred for a description of this method for estimating the number of statistically significant latent roots in geologically relevant material. The rule of thumb is to accept significance if the value of PRESS is larger than one. In the present case there are four values greater than unity after which there is a rapid fall-off. Using this information, the analysis was pursued using the information residing in the first four principal components. PRESS is defined as:

$$\mathrm{PRESS}(m) = 1/np\ \Sigma_{i=1,p}\Sigma_{j=1,p}(\sim x_{ij}{}^{(m)} - x_{ij})^2,$$

where m is the number of components, p is the number of variables, $\sim x$ denotes predicted values, and n is the number of individuals in the sample.

The increase in predictive information W_m supplied by the mth component is computed as:

$$W_m = ([\mathrm{PRESS}(m-1) - \mathrm{PRESS}(m)]/D_m)/(\mathrm{PRESS}(m)/D_r),$$

where D_m is the number of degrees of freedom needed to fit the mth component and D_r is the number of degrees of freedom remaining after fitting the mth component.

Cross-Validation Principal Component Analysis for Ten Variables

The plot for the first two principal component scores for the correlation matrix of the measurements (Fig. 6.3) shows that there is a zone of overlap between juvenile conchs and microconchs. Apart from one specimen of macroconchs (specimen 13), macroconchs are efficiently separated.

Redundant Variables

Table 6.1 summarizes the results for assessing redundant variables, computed for the entire sample of 25 specimens. These results indicate that the variables – ventral configuration (var 2), penultimate ventral width (var 6), location of the umbilicus (var 8), and umbilical width (var 9) perturb the analysis if removed and may be worth further consideration in interpreting the variability in the dimensions of the conchs. The entries in bold type denote variables that do not contribute major information to the analysis. The set of observations obtained from the examination of redundancy and information (Table 6.1) encompasses the four distances 2, 6, 8, 9. Thus, two apertural and two lateral distances stand forth as being the most

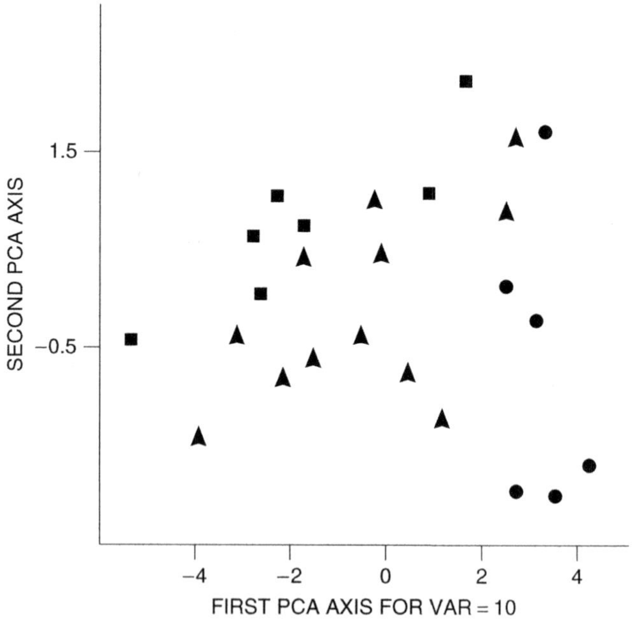

Fig. 6.3 Plot of the first two principal component scores (10 variables). Key to symbols: *Squares* = points for juvenile conchs (N = 7), *Dots* = points denoting macroconchs, (N = 6) *Arrow-heads* = points for microconchs (N = 12)

Table 6.1 Identification of redundant variables by deleting each variable in turn from the analysis: residual sums of squares for the Procrustean fit of new scores to old scores

Variables removed from the analysis (with subsequent replacement)	Principal component spaces examined			
	P1	P1 + P2	P1 + P2 + P3	P1 + P2 + P3 + P4
1	1.446	1.539	1.532	1.538
2	1.248	4.020	**6.707**	**6.578**
3	1.298	1.456	1.556	2.068
4	1.302	1.421	1.448	1.571
5	1.240	1.251	1.258	1.279
6	1.227	4.732	**5.591**	**6.020**
7	1.390	1.686	1.676	1.690
8	1.316	1.860	**5.411**	4.451
9	1.305	2.000	**6.890**	**5.200**
10	1.421	1.417	1.380	1.379

Note: Bold entries denote deletions that strongly perturb the principal component residuals

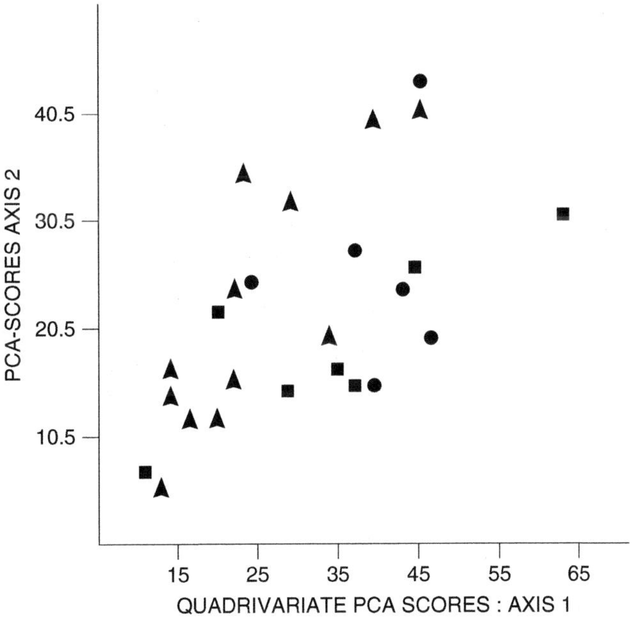

Fig. 6.4 Plot of the quadrivariate set of the first two principal component scores. Symbols the same as in Fig. 6.3

important for establishing differences between individuals of the three categories involved. The plot of the first two principal components displayed in Fig. 6.4, displays a less marked differentiation between categories. Several macroconchs now fall within the microconch field and the differentiation between juveniles and macroconchs is not clearly maintained The analysis yields the informative result that the dispersion of points in the plot (Fig. 6.4) is somewhat less heterogeneous than for the full set of data. How is this result to be interpreted? One logical explanation would seem to be that the four dimensions maximum ventral width, penultimate ventral width, umbilical location and width of the umbilicus are inherently stable properties of the species, shared by most of the individuals available for study, and hence not affected to a marked extent by dimorphism or by environmental pressures. Fig. 6.3 shows also that juveniles are spread over the first axis whereas macroconchs are spread in the direction of the second axis, as are also the microconchs. This could imply that there is room for error in identifying the three categories. A possible source of interference could be caused by the effects on buoyancy of the conch occasioned by epibionts settling during the life of the animal (Seilacher 1960; Meischner 1968). The added weight of a ballast of oyster encrustations and barnacles could trigger the sutural adjustment biomechanism for achieving neutral buoyancy of the animal before maturity had been reached, thus bringing about the appearance of sutural crowding by an external source not due to impending sexual maturity and the cessation of growth. A further possible factor could be that the informative four variables are not linked to sex and ontogeny as markedly as in the full analysis but rather to such properties as egression of the umbilicus and other features reflecting maturity (Fig. 6.2b illustrates the manifestation of egression of the umbilicus).

Statistical Distances Between Individuals

The Q-mode method of principal coordinate analysis (Gower 1966) is of great graphical value in studies of morphological variability in various organisms, invertebrates as also vertebrates. An analysis of the full data set is displayed in the graph in Fig. 6.5. This shows that macroconchs are dispersed along the second coordinate axis, with little lateral displacement. Microconchs and juveniles are spread diagonally along both axes in a manner such as to suggest that juveniles and microconchs are in some manner bound to each other in a morphological sense. This echoes the result yielded by the principal components. The significance of this result is that there could be a confusion in the assignation of some juvenile specimens, and, or, vice versa. This possible source of inaccuracy is not manifested in the specimens identified as macroconchs. The figures located alongside the plotted points, and enclosed in brackets, serve to give an idea of the general shape of the conchs. These figures denote the ratio between the diameter of the conch and its maximum thickness thus providing a simple expression of the apertural "aspect-ratio" of the shell. It is noteworthy that there is no hard and fast separation connecting to inflation, although there is a tendency for larger values to lie in the upper third of the graph and ordinated along the second principal coordinate axis. The percentage residual for the first two latent roots of the principal coordinates is 47.1%, which in the distance-preserving connexion of principal coordinate ordination represents a reliable result.

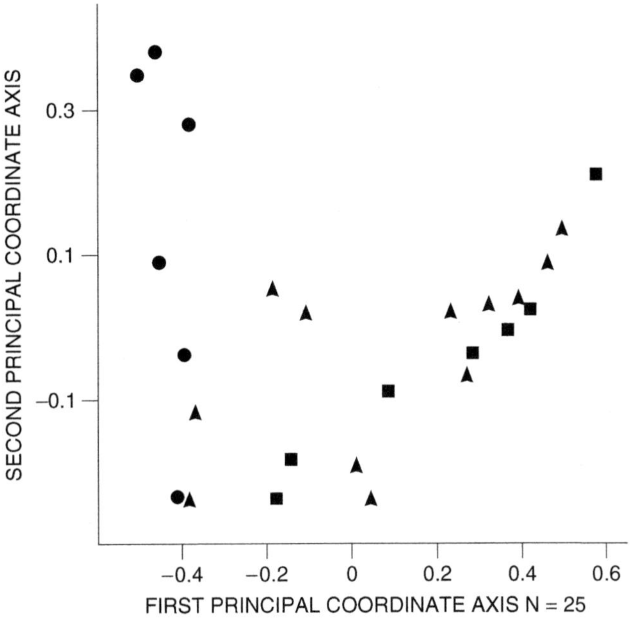

Fig. 6.5 Principal coordinates plot for the first two axes. Symbols the same as in Fig. 6.3

Table 6.2 Comparisons between latent roots for macroconchs, microconchs and juveniles (principal components computed for correlations)

Macroconchs versus microconchs			
Latent root 1	7.43°	Latent root 2	69.52°
Macroconchs versus juveniles			
Latent root 1	70.83°	Latent root 2	55.46°
Microconchs versus juveniles			
Latent root 1	70.34°	Latent root 2	66.12°

Angles Between Latent Vectors

The angles computed between the first and second latent vectors (principal components) for macroconchs, microconchs and juveniles are listed in Table 6.2. There are several features worth commenting upon. The first latent roots for macroconchs and macroconchs indicate that the major ellipsoidal axes are almost identically oriented. However, the angle between the second ellipsoidal axes is markedly different. The comparisons for macroconchs versus juveniles and microconchs versus juveniles are all quite different. This information discloses that the three categories are statistically very unlike on the whole, which undoubtedly lends confidence to the criteria used for characterizing them. These results indicate that the covariance matrices of the three samples are markedly different which prohibits the possibility of using standard procedures of multivariate analysis and morphometrics (cf. Hoff 2009). This heterogeneity in variability patterns is connected with the morphological variability within the species.

Critical Angles Between Specimens

The analytical procedure requires the computation of *critical angles*, that is, a measure of influence of each individual in the sample, expressed as $t = \text{arcos}(\vartheta)$ where ϑ denotes the smallest element of the diagonal matrix **L** defined above. Large angles are associated with *influential* observations in the sample (Krzanowski 1987b, pp. 167–176). The maximum critical angle is a convenient and suitable measure of influence of each specimen in the sample. The greater the angle the greater is the perturbation to the principal component brought about by the deletion of the corresponding specimen and hence an indicator of its relative importance.

Table 6.3 Maximum critical angles obtained from the successive deletion of specimens with replacement

Removed specimen		Largest principal components (variance sensitive)				Smallest principal components (correlation sensitive)			
		1	2	3	4	−4	−3	−2	−1
1	J	0.786	1.558	1.811	3.204	**17.240**	9.796	6.440	2.292
2	J	1.472	3.733	4.296	2.563	4.098	1.696	1.503	1.510
3	J	0.623	4.777	6.600	5.308	7.409	9.959	**21.908**	5.491
4	J	1.206	1.972	1.819	1.819	6.886	1.273	0.169	0.007
5	J	1.776	**11.280**	**12.066**	**29.315**	**50.844**	**10.917**	3.082	2.959
6	J	1.946	1.609	1.650	2.389	5.100	3.081	4.174	1.957
7	J	0.248	0.966	2.896	6.148	1.683	5.535	5.681	3.747
8	MA	1.077	1.159	1.565	1.790	**15.055**	2.081	1.705	1.589
9	MA	1.252	1.315	1.503	4.671	**17.100**	7.922	4.522	4.324
10	MA	0.981	1.733	2.367	2.108	2.115	2.093	3.051	3.294
11	MA	2.935	**11.105**	**12.066**	**11.211**	**16.359**	**41.996**	**72.317**	6.748
12	MA	1.095	1.161	5.392	4.806	**10.651**	5.629	6.293	5.135
13	MA	4.828	**11.581**	**16.899**	**11.615**	**12.302**	**20.230**	**42.039**	**19.744**
14	MI	3.662	9.528	9.260	4.253	5.687	3.986	5.097	1.061
15	MI	3.074	6.964	6.981	**14.997**	**31.292**	9.026	9.352	0.890
16	MI	1.051	1.066	2.654	2.992	**51.860**	7.013	7.011	4.606
17	MI	0.149	2.314	2.316	5.057	9.073	6.775	7.149	4.320
18	MI	2.079	3.288	2.981	2.973	6.795	4.991	6.291	6.343
19	MI	0.097	1.911	7.902	8.475	**59.384**	**18.582**	**18.131**	**18.614**
20	MI	0.877	2.309	5.291	5.490	**39.312**	6.245	1.990	1.989
21	MI	2.983	3.231	2.874	2.885	6.877	5.961	7.016	**29.343**
22	MI	0.222	2.377	3.909	3.938	**13.476**	**11.061**	**13.866**	**10.060**
23	MI	0.776	3.713	3.944	8.226	**11.539**	9.565	**16.325**	**13.341**
24	MI	1.964	5.033	**10.692**	**10.017**	**51.798**	7.310	1.706	0.231
25	MI	0.767	1.302	2.873	3.169	**24.392**	7.996	8.313	7.257

Deletion of specimens with subsequent replacement
J juvenile, *MA* macroconch, *MI* microconch

Table 6.3 summarizes the results of the estimation the influence of the i-th specimen on the analysis. Cross-validation has shown that the functional dimensionality of the problem is four. Table 6.3 lists the largest critical angle between the plane defined by the first four principal components computed from the full sample and the plane defined by deleting each sample in turn. This information is supplied in the column headed with a "2". The columns headed with "3" and "4" list the critical angles for three- and four-dimensional spaces. The largest values in these columns betoken individuals, the removal of which cause the greatest disturbances in the principal component analysis. These are indicators of outliers of location or dispersion. The columns headed by negative numerals define the planes of the smallest principal components. The smallest principal components reflect outlying observations produced by deviating correlations. There is no test available for assessing the criticality of an angle; a reasonable guideline is to accept values of ten or greater as being indicative of significance (Krzanowski 1987b). The results listed in Table 6.3 show that there are some very large differences, the most notable of which are for a juvenile conch (specimen 5), a macroconch (specimen 11) and several microconchs (specimens 13, 15, 19, 22, 23, and 24). We note that some specimens are aberrant both with respect to the largest principal components as to the smallest. These are 5 (juvenile), 11 (macroconch),13 (microconch), and 24 (microconch).

Discussion

In many organisms polymorphism may be of several kinds and that within the same species discontinuous variations in size, shape and ornament. Levinton (1988) has devoted considerable attention to aspects of this subject. Reyment and Kennedy (1991, p. 417) summarized work on gastropods which they found to be of relevance for interpreting variability in ammonite cephalopods, including important results of Goodfriend (1986), Via and Lande (1985), and Bulmer (1980). The analysis of variability in the ammonite species reported on here raise several items of interest. These are:

1. Does the analysis support accepted dogma for sexual dimorphism in the proportions of the conch. By and large this seems to be the case, but there is a lack of consistency with respect to the great variability in the shell morphology of a part of the material. This aspect has been exhaustively documented iconographically in the monograph by Kennedy et al. (2009). There is, however, a problem in need of elucidation, notably, the fact that most ammonite species do not show dimorphic properties, nor polymorphism of other kinds and it seems possible that instability in shape and ornament is connected to a shallow water environment. This supports the hypothesis of Davis et al. (1996).
2. At the level of morphological integration (Olson and Miller 1958) an unawaited result is that the correlations are very unlike. The best agreement is for the angle between the first latent vectors of the hypersphere for macroconchs versus

microconchs. The angles between the first latent vectors between microconchs and macroconchs and juveniles are large. The angles between the second latent vectors of all three categories are likewise large. The strongly expressed differences in morphological integration between categories of conchs is puzzling. On present knowledge, however, there is no logical reason for reverting to the solution of Collignon (1983) of referring each morphology to a separate species.

References

Aitchison, J. 1986. The statistical analysis of compositional data. Chapman and Hall, London.

Barber, W. M. 1957. The Lower Turonian ammonites of northeastern Nigeria. Bull. Geol. Survey Nigeria, 26: 1–86.

Blackith, R, E. 1965. Morphometrics. In "Theoretical and Mathematical Biology" Eds. T. H. Waterman and H. J. Horowitz, pp. 225–249. Blaisdell Publishing Company, New York.

Bulmer, M. G. 1980. The Mathematical Theory of Quantitative Genetics. Oxford University Press, Oxford.

Callomon, J. H. 1963. Sexual dimorphism in Jurassic ammonites. Transactions of the Leicester lierary and philosophical Society, 57: 21–56.

Collignon, M. 1983. Faune albo-cénomanienne de la formation des marnes de Kazhdumi, region de Fars-Kuzhestan (Iran). Doc. Lab. Géol. Lyon (hors série), 6: 252–291.

Davis, R. A., Landman, N. H., Dommergues, J. I., Marchand, D., Bucher, H. 1996. Mature modifications and dimorphism in ammonoid cephalopods. In: Landman, N. H., Tanabe, K., Davis R. A. (Eds.). Ammonoid Paleobiology. Topics in Geobiology vol. 13, 463–539.

Gower, J. C. 1966. Some distance properties of latent roots and vectors methods used in multivariate analysis. Biometrika, 53: 325–338.

Gower, J. C. 1975. Generalized Procrustes Analysis. Psychometrika, 40: 33–57.

Goodfriend, G. 1986. Variation in land-snail shell forms and its causes: a review. Systematic Zoology, 35: 204–223.

Hoff, P.D. 2009. A hierarchical eigenmodel for pooled covariance matrices. J. R. Statistical Society B, 71: 971–992.

Kennedy, W. J. , Reyment, R. A., MacLeod, N., Krieger, 2009. Species discrimination in the ammonite genus *Knemiceras* Böhm. Palaeontographica Abteilung A, 290, 1–63.

Krzanowski, W. J. 1982. Between-group comparison of principal components – some sampling results. Journal of Statistical Computing and Simulations. 15: 141–154.

Krzanowski, W. J. 1987a. Selection of variables to preserve multivariate data structure using principal components. Applied Statistics, 36: 22–33.

Krzanowski, W. J. 1987b. Cross-validation in principal component analysis. Biometrics, 43: 575–584.

Levinton, J. 1988. Genetics, Paleontology and Macroevolution. Cambridge University Press, New York.

Makowski, R. 1963. Problem of the sexual dimorphism in ammonites. Paleontologica Polonica, 12: 1–92.

Meischner, D. 1968. Perniciöse Epökie von *Placunopsis* auf *Ceratites*. Lethaia, 1: 156–174.

Mutvei, H. and Dunca, E. 2007. Connecting ring ultrastructures in the Jurassic ammonite *Quenstedtoceras* With discussion on mode of life of ammonites. In Landman et al. (Eds.), Cephalopods Past and Present: 239–256, Springer Verlag.

Olson, E. C. and Miller, R. L. 1958. Morphological Integration. Chicago University Press.

Reyment, R. A. 1971. Vermuteter Dimorphismus bei der Ammonitengattung *Benueites*. Bull. Geol. Instn Univ Uppsala, N.S. 3: 1–18.

Reyment, R. A. 1988. Does sexual dimorphism occur in Upper Cretaceous ammonites. Senckenbergiana Lethaea, 69: 109–119.

Reyment, R. A. 1991. Multidimensional Palaeobiology. Pergamon Press, Oxford, 377 + 39.

Reyment, R. A. 2003. Morphometric analysis of variability in the shell of some Nigerian (Turonian) ammonites. Cretaceous Research, 24: 789–803.

Reyment, R. A. 2004. Instability in principal component analysis and the quantification of polyphenism in palaeontological data. Math. Geol., 36: 629–638.

Reyment, R. A. and Jöreskog, K. G. 1993. Applied factor analysis in the Natural Sciences. Cambridge University Press, 371.

Reyment, R. A. and Kennedy, W. J. 1991. Phenotypic plasticity in a Cretaceous ammonite analysed by multivariate statistical methods. Evolutionary Biology, 25: 411–426.

Reyment, R. A. and Savazzi, E. 1999. Aspects of Multivariate Analysis in Geology. Elsevier, 284.

Seilacher, A. 1960. Epizoans as a key to ammonite ecology. Journal of Paleontology, 34: 189–193.

Swan, R. H. and Saunders, W. B. 1987. Function and shape in Late Paleozoic (mid-Carboniferous) ammonoids. Paleobiology, 13: 297–311.

Urlichs, M. 2006. Dimorphismus bei *Ceratites* aus dem Germanischen Oberen Muschelkalk (Ammonoidea, Mitteltrias) mit Revision einiger Arten. Stuttg. Beitr. Naturk., Ser. B, 363, 1–85.

Urlichs, M. 2009. Weiteres über Dimorphismus bei *Ceratites* (Ammonoidea) aus dem Germanischen oberen Muschelkalk (Mitteltrias) mit Revision einiger Arten. Neues Jahrbuch Geol. Paläont. Abh. 25: 199–223.

Via, S. And Lande, R. 1985. Genotype-environment interaction and the evolution of phenotypic plasticity. Evolution, 39: 297–311.

Wold, S. 1978.Cross-validatory estimation of the number of components in factor and principal component models. Technometrics, 20:397–405.

Chapter 7
Development and Applications of Computed Tomography in the Study of Human Fossil Crania

Eva María Poza-Rey and Juan Luis Arsuaga

Introduction

Radiographic techniques have been used in paleoanthropology and comparative anatomy since the discovery of X-rays (Branco 1906). Each advance in medical imaging technology has been accompanied by new applications in both fields (Mafart et al. 2004).

The techonolgy to develop X-rays was discovered on November 8, 1895, and the first application to human beings was made in Germany by the physicist Wilhelm Conrad Röntgen. The first photograph of Röntgen's wife's hand was taken on December 22, 1895.

This opened a full range of advantages to the medical world, especially in orthopaedic surgery, as the most obvious clinical application of the X-ray was for detection and characterization of fractures and dislocations (Van Tiggelen 2001). Another important application was the localization of foreign bodies, such as bullets (Davidson 1916).

At this time, the concept of the evolutionary origin of the human species, i.e., the theory that man and the great apes shared a common ancestor in the past, was well accepted among scientists.

Soon after the discovery of X-rays, radiography was applied to newly found hominid fossils, such as the Krapina Neanderthal remains (Gorjanovic-Kramberger 1906; and also see Zonneveld et al. 1989; Weber et al. 2001; Tobias 2001;

E.M. Poza-Rey (✉)
Centro Mixto UCM-ISCIII de Investigación sobre Evolución y Comportamiento Humanos, Instituto de Salud Carlos III, c/Monforte de Lemos 3, 28029 Madrid, Spain
e-mail: epoza@isciii.es

J.L. Arsuaga
Centro Mixto UCM-ISCIII de Investigación sobre Evolución y Comportamiento Humanos, Instituto de Salud Carlos III, c/Monforte de Lemos 3, 28029 Madrid, Spain
and
Departamento de Paleontología, Facultad de Ciencias Geológicas, Universidad Complutense de Madrid, Ciudad Universitaria s/n, 28040 Madrid, Spain

A.M.T. Elewa (ed.), *Computational Paleontology*,
DOI 10.1007/978-3-642-16271-8_7,

Wilhelm Conrad Röntgen

Mafart et al. 2004) and the Mauer mandible (Schoetensack 1908). Although at first it was mainly used to study roots of the teeth in the jaw, it was also later applied to complete skulls in order to visualize the pneumatisation in *Homo erectus* found in China in the 1920s and 1930s of the twentieth century (Zonneveld et al. 1989).

During World War I, training programs for radiologists and technologists were required, and portable X-ray units were developed. X-ray pictures were two-dimensional representations of three-dimensional objects. This was a problem, and an attempt to solve it was by obtaining images in various projections, but this caused an overlap of the structures of the skeleton and other organs. Furthermore, mineralization or the presence of sedimentary matrix, the small size of the intracranial structures, and the superimposition of structures were also limitations for the study of fossilized human bones (Wind and Zonneveld 1989; Van Tiggelen 2001; Mafart et al. 2004).

A new method was developed to solve these limitations of the conventional radiology. This new method is known as tomography, and was discovered during World War I in 1916. One of the names that contributed to this discovery, is that of a French dermatologist, André Bocage, who described the method in 1921 in *Procede et dispositifs de radiographie sur plaque en mouvement* (*French patent 536464*) (Van Tiggelen 2001; Goldman 2007). Other names that claim to be the inventor are: Ziedses des Plantes, Bartelink and Vallebona (Franz W. Zonneveld pers. comm.).

Conventional tomography allows radiological cross-sections to be created of an organ at any given depth without superposition of anatomic structures. However, the most important technological progress was made with the use of computers for decoding crypted military messages from the German and Japanese armed forces during World War II. This quickly resulted in numerous applications in both military and civilian domains. Hence, the axial computer-assisted tomography was achieved, based on the principles of tomography discovered in the First World War and the power of the computers developed in the Second World War (Van Tiggelen 2001).

From April 1972, when CT was introduced, there have been important improvements in the technique, such as beam-hardening correction in 1975 (Kijewski et al. 1978; Jian and Hongnian 2006) or breath-hold scanning times in 1976 (Raptopoulos et al. 1978; Goodman 2010).

Godfrey N. Hounsfield

Allan M. Cormack

But the greatest advancement of computed tomography, was recognized in the late 1970s to **Godfrey N. Hounsfield and Allan M. Cormack**, who were awarded with the Nobel Prize for Medicine in 1979, *"for the development of computer assisted tomography"*.

Just a few years later, and despite the limitations of the radiography, this led to extensive work carried out by Skinner and Sperber (1982). It consisted of the most comprehensive collection of radiographs of fossil hominids ever assembled, published as *The Atlas of Radiographs of Early Man* (Tobias 2001). Although previous studies do exist, such as the one made for a pioneer using radiography in fossil hominids studies, Franz Weidenreich (1943).

During the 1980s, a group of scientists started to apply the advances in 2D and 3D computed tomography (CT) to look inside fossil long bones (Jungers and Minns 1979; Tate and Cann 1982) and fossil skulls (Wind 1984; Laitman 2004). The fast development that medical imaging underwent during that decade, was smartly applied to paleontology, and CT became a very useful instrument in the study of the fossil hominins (Wind 1984; Wind and Zonneveld 1985, 1989; Zonneveld and Wind 1985; Zonneveld et al. 1989; Mafart et al. 2004).

Conroy and Vannier (1984) were the first to use 3D CT in the field of paleontology, for the study of the endocranial surface of an Oligocene ungulate cranium (Spoor and Zonneveld 1999). Since then, this technique has been applied in paleoanthropology for several purposes.

The reconstruction of three-dimensional images from sequential series of computed tomography scans, has become a common technique in medical practices (Hemmy et al. 1994; Zonneveld and Fukuta 1994; Zonneveld 1994). The advances in the method are now enabling a level of exploration, visualization, assessment, and quantification that was previously unattainable (Laitman 2004).

Thanks to development of this methodology, nowadays, the potential study of fossils with the application of CT and three-dimensional analysis, allows for multiple studies of endocranial regions, as well as: reconstruction of brain endocasts, sinus cavities, reconstitution of incomplete crania, release of material embedded in soil matrix and virtual manipulation of fossils. Furthermore, the reconstruction of functional characteristics of the hominids is possible, such as those described in the first computerized analysis of STS 5 (Wind and Zonneveld 1989):

> the assessment of the skull base anatomy may assist in reconstructing the pharyngeal shape and laryngeal position, and hence ancestral vocal abilities; the knowledge of the intratemporal structures may provide indications for the original position of the head in space (semicircular canal position) (Spoor and Zonneveld 1998) facial mobility (facial nerve canal size could provide indications for the nerve size); acoustic properties (external, middle and inner ear morphology).

All these possibilities, increase the knowledge and study of the fossils, protect their integrity and preserve their conservation status.

What Is Computed Tomography?

Conventional radiology, first used in the study of fossil remains (see "Introduction" section), had many limitations when applied to the study of fossil material.

Goldman (2007) exhibits some of these limitations, briefly revised as follows:

- *Inefficient X-ray absorption*: The film-screen used 20–25 years ago, had an X-ray absorption efficiency of 25%. In this way, 75% of the energy was lost so information is affected too.
- *High Scatter to primary X-ray ratios*: If 50% of the detected X-rays are scatter, then subject contrast is reduced by a contrast reduction factor of 0.5.
- *Superimposition and conspicuity*: Conspicuity is the ease of finding an image feature during a visual search. 3D volume is reduced in a 2D image, so underlying tissues and structures are superimposed and result in reduced conspicuity.
- *Receptor contrast vs. latitude*: Radiographic films must provide sufficient exposure latitude to record as much of the range of X-ray intensities exiting the scanned object; this feature necessarily limits receptor contrast. Modern radiographic technology and digital radiography have improved X-ray absorption efficiency, but most of the limitations of radiography due to contrast-latitude still exist today.

Brief Historical Context

Computed tomography, was developed as a method to solve part of the limitations mentioned previously, and was a revolution in radiology.

The theory of image reconstruction from projections, which is central to the basic concept of CT, was discovered in 1916 by André Bocage, described in 1917 and patented in 1921/1922. Another person that has to be considered too is the Austrian mathematician Radon (1917).

Subsequent patents and new development of the method occurred from 1937 onwards and was proposed for medical imaging as early as 1940 (Webb 1992; Van Tiggelen 2001; Van Tiggelen and Pouders 2003; Goldman 2007).

During the 1950s, the experimental scanner, medical scanner CT, transmission CT, and emission CT were developed and tested for the improvement of the method. The development of the first modern CT scanner was finally begun in 1967 by Sir Godfrey Newbold Hounsfield, an engineer at British EMI Corporation, in collaboration with the Department of Health and Social Security and Atkinson Morley's Hospital (London), and was commercially available in 1972 (Webb 1992; Goldman 2007).

For more details, dates and names, see *From the Watching of Shadows: the Origins of Radiological Tomography* by Steve Webb (1990) and the historical articles published by Van Tiggelen (2002) and Van Tiggelen and Pouders (2003).

An extensive description about development of CT, from the first to latest generations, can be found in Goldman (2007).

Definitions and Concepts of the Method

CT is defined as a radiographic technique for diagnostic imaging, that uses a combination of X-rays and computer technology to assimilate multiple X-ray pictures and produce cross-sectional images (slices), both horizontally and vertically of a body. The vertical images are derived from the horizontal ones by a calculation process called multiplanar reformatting.

CT scans are more detailed in terms of contrast resolution than general X-rays, and do not suffer from the superimpositioning of structures outside the plane of interest. This quality allows CT scans to visualize small density differences (Zonneveld and

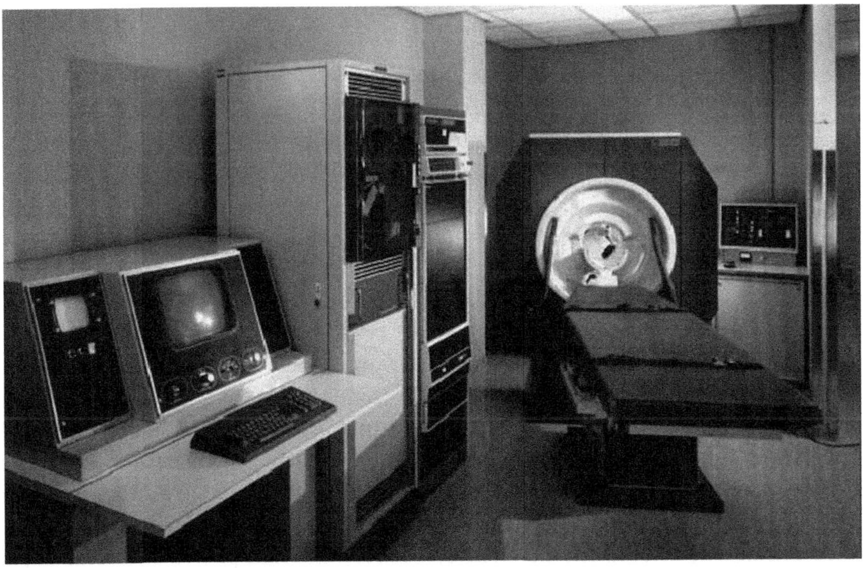

A second generation EMI scanner, called CT 1010. Extracted from Van Tiggelen (2001). The First CT scanner in continental Europe was the first generation EMI Mark I CT-scanner in the Karolinska Hospital in Stockholm, Sweden

Wind 1985; Wind and Zonneveld 1985). CT is compounded by an X-ray generation system, a data collection system and a data processing system. The X-ray generation system contains the high-tension generator, an X-ray tube and some tube collimators. The data collector system, is made up of a detector system, a data acquisition system, an analog-to-digital converter and a calibration and transmission system for the data.

Jiménez-Castellanos (1981) summarizes the process in a manual on computed tomography as follows: *an X-ray tube that emits a photon beam, which, after passing through the explored matter, is collected after attenuation by a detector system, which sends the information to a computer (data acquisition system), responsible of complex mathematical operations and transform the data into an image yielded by a recorder device.*

The CT computer system, amplifies the signal generated by the detector, and converts the analogue signal in digital information.

Basic characteristics of the CT images are determined by a few factors; the image is captured by sensors connected to a computer, and the radiation-emitting tube does not remain static as in the radiograph (which produces a "snapshot" flat image as if it were a photograph) but moves around the area of interest. This makes the scanned area become a volume consisting of an array of elementary volumes called voxels.

In comparison, the digital radiographic image is a flat image that consists of a minimum of surface units called pixels (Arana-Fdez. de Moya et al. 2006). Also, Doyon et al. (1995) states, according to figures measured by peripheral absorption detectors, the computer calculates the different densities encountered by the radiation per unit of volume, which we call voxel, and then the pixel is defined as the surface of the base of the voxel.

The pixel matrix commonly used in Computed Tomography, vary from 256 × 256, 512 × 512 or 1,024 × 1,024, and this gives a voxel quantity of 65,536; 262,144; and 1,048,576 respectively, which gives a better image definition.

The slice thickness indicates how thick, in millimeters, your slices are and is equal to the height of the voxel. In the first generation of CT, the width of the X-ray beam specifies the slice thickness to be imaged (Goldman 2007). First radiological applications on fossils, used medical tomographs with limitations, since the slice thickness is high, with values from 8–10 mm in first equipments (Zonneveld and Wind 1985; Le Floch-Prigent 1989) to 0.5 mm in most modern one. Along time, and with the development of industrial tomography and microtomography, the slice thickness has decreased to 0.01 mm and 200 μm respectively (Spoor et al. 2000a; Mafart et al. 2004; Mazurier et al. 2006). So, definition of subsequent virtual fossil reconstruction has been improved, although time of exposure has been increased too. However, this parameter does not affect the integrity of fossils.

The slice index, also called slice interval, slice distance (or slice overlap, which is a secondary parameter that occurs as soon as the slice index is smaller than the slice thickness), determines the distance between the mid-planes of two adjacent slices. It is possible to use a slice index larger than the slice thickness, but this will leave some areas unexplored, or one can use overlapping slices. As an example, a slice thickness of 5 mm, with a slice index of 5 mm, results in a study with contiguous slices without unexplored areas. With the advent of spiral CT-scanning (since 1989) overlapping slices are the modus operandi, as this produces better multiplanar reformats and better 3D images while it doesn't cost more radiation, though that is irrelevant for fossil scanning (Franz W. Zonneveld pers. comm.).

A slice thickness of 1.5 mm, with a slice index of 1 mm, would result in a study with overlapping images, which allows for a subsequently better 3D reconstruction.

F.O.V. (Field of view) is the size of the image reconstruction. The area encompassed by the X-ray acquisition is called "Scanned area". FOV can be the same as Scanned area or smaller. In the latter case one speaks of a Zoom reconstruction (Franz W. Zonneveld pers. comm.).

The values can be too variable, and are adjusted to include the entire subject to be scanned (MacLeod et al. 2003) so the diameter of the slice is determined. As the FOV increases, the matrix size remains the same, and the pixel size will be increase (Spoor et al. 2000a).

By definition, as it appears in Jeffery and Spoor (2002), pixel sizes for each reconstruction can be computed by dividing the field of view (FOV) by the reconstruction matrix size.

kV and mA. These are the dimensions in which "Tube voltage" and "Tube current" are represented. These parameters are related in the scanning process. High kV values, require low values for mA, but there is a complex relationship between both parameters depending on the loadability of the X-ray tube. They correspond to the shot scanning features, and they usually are registered for each type of exploration in the scanner, but the parameters can be changed manually. Adequate technical adjustment (as reduction of the tube current) can reduce the radiation with a high percentage without loosing image quality (Arana-Fdez. de Moya et al. 2006).

The lower the tube current (mA) and exposure time are, the lower the resulting radiation dose will be. Low tube voltage, results in low-contrast detectability (Funama et al. 2005) but increases the dose for the same detector signal.

A disadvantage of the CT technique with low tube voltage is the increase in noise. As Funama et al. (2005) indicates in their work, Boone et al (2003) found a relationship between image noise, the tube voltage, and the tube current–time product setting in CT. They showed that noise increased at lower tube current–time product settings and lower tube voltage.

Radiation dosage, does not affect the fossil integrity, as it does in alive patients, and that is why medical scanners work with low tube voltage (low values for kV).

> Sometimes fluorescence is used to measure the age of a fossil (the amount of cosmic rays captured by the fossil is a measure for time and is converted to light during a heating process), however, the amount of radiation used by CT-scanning is usually much smaller than the amount of cosmic rays the fossil has absorbed over its time of existence (Franz W. Zonneveld pers. comm.).

Another way to reduce the radiation dose, is to restrict the explored area (Arana-Fdez. de Moya et al. 2006). This appears as a limitation for the study of fossil remains, e.g., in situations of virtual extraction of rock matrix and the imaging and isolation of small structures.

To solve these restrictions, industrial- and micro-CT scanners (high resolution and ultra-high resolution) have been developed and extensively used over the last years. This type of scans, allow for an incredible level of contrast detectability. Contrary, low values of mAs appear associated (Zollikofer et al. 2005; Bush et al. 2004; Bastir and Rosas 2005; Kuroe et al. 2004; Bastir et al. 2004; Ketcham and Carlson 2001; Bush 2004; Thompson and Illerhaus 1998). As the radiation dose is linear with the mAs when all other factors are held constant, then constant radiation doses will show low values too.

Virtual Anthropology (VA) and Computer-Assisted Paleoanthropology (CAP)

Application of the Medical-Engineering Technologies to the Study of Human Fossils

Virtual Anthropology provides procedures to investigate three-dimensional morphological structures by means of digital data sets of fossil and modern hominoids within a computer environment (Weber et al. 2001).

Data can be acquired by different processes, and it depends of the analysis requirements. Here is a description of the most used methods (based on Calhoun et al. 1999; Weber 2001; Wilhite 2003; the specific Ph.D. of Neeser 2007).

- *Surface*: Laser-scanner is the method usually considered for surfaces. The advantage of the technique is that the laser scanner requires no contact with the fossil specimen. The posterior surface rendering, includes speed and flexibility in image rendering, and permits detailed 3D data to be gathered from complex joint surfaces (Aiello et al. 1998).

Triangulation scanners appear to be of the most used for acquiring models from fossil specimens.

> *Active triangulation scanners*. These operate by passing a stripe (drawn by laser) down the object's surface. A camera, typically a CCD, measures the stripe's displacement, and in so doing determines surface coordinates. Triangulation scanners are typically used for close range work, below two meters, with possible accuracies of micrometres (Schulz and Ingensand 2004). For example, Minolta's VIVID 9103 laser scanner has an effective range of between 0.6 and 2.5 m, and a reported accuracy of 8 μm (Neeser 2007).

Some years before, Aiello et al. (1998) already achieved an accuracy of 0.1 mm in a scanned talocrural joint surface.

For more information of Laser range scanning methods, see Chap. 3 of "A Comparison of Statistical and Geometric Reconstruction Techniques: Guidelines for Correcting Fossil Hominin Crania. PhD Thesis. Neeser 2007".

- *Landmarks or contours*: There are two types of commonly used contact digitizers to take this data:
  1. *Mechanical digitizers* measure the angles between various segments of a mechanical armature supporting the digitising pen. Using both the angles and the known segment lengths, the position of the pen tip can be calculated.
  2. *Magnetic digitizers,* based on a magnetic field dependent system, it measures the position of a magnetised tip within a larger magnetic field.

Digitized data is very useful, because it can be quickly captured by the researcher, the contact digitiser is relatively easy to use, it is cost efficient, and consisted of relatively small digital files that are easily manipulated on a standard laptop or desktop computer. When compared to many of the other imaging modalities, especially CT, the cost of obtaining a contact digitiser is rather minimal and their portability makes it easy to take the digitizers to the specimen, rather than vice versa. The benefits of digitized data make it especially useful in morphometric, ontogenetic, and biomechanical studies. However, potential problems, such as extra time needed to assemble skeletal elements and accuracy limitations of digitized data, should be considered when contemplating using a three-dimensional digitizer to capture morphological data.

- *Volume*: medical diagnostic radiology (Computed Tomography, CT). You obtain an almost exact virtual copy of the original. Some limitations such as the resolution, which is limited, and the parameter imaged is an X-ray attenuation coefficient which may be indiscriminate when trying to separate certain structures from one another.

This methodology, together with a powerful software to visualize the virtual object, allows for the most complete studies and manipulations, as you can scale, magnify, rotate, cut, move, measure and image. This method is the basis for the Computer-Assisted Anthropology/Paleoanthropology (CAA/CAP).

The quality of data generated from modern medical scanners continues to improve. The evolution from conventional tomography to the latest CT-scanner generation has brought advantages for 3D imaging. Newer scanners and recent types of CT technology (microtomography, nanotomography...) allow longer acquisition times, resulting in larger volumes of very high resolution data and quality.

This kind of data acquisition is expensive, in equipment and time, but the high resolution data are ideal for 3D imaging.

The development of 3D imaging itself has a long history. First there was the depth-encoding technique with its derivative gradient shading [as used by Conroy and Vannier (1984)]. Then, there was the surface rendering technique that took the orientation of the surface into account, then there was the volume rendering technique, and finally a hybrid version of the latter two techniques resulting in the sophisticated images with an opaque brain cavity and a transparent skull of the LB1 *Homo floresiensis* skull (Franz W. Zonneveld pers. comm.).

Weber (2001) determines the importance of these techniques in the study of fossils, revealing four important points:

1. The accessibility of all (including hidden) structures
2. The permanent availability of the virtual objects
3. The general accuracy and reproducibility of measurements
4. The possibility to obtain information for advanced morphometric analysis

Furthermore, CT-scanned fossils, can be subsequently recreated physically, by rapid prototyping using techniques such as **stereolithography**: a 3D layering process, which allows to create physical three-dimensional (3D) objects from CAD images (see Seidler et al. 1997; Recheis et al. 1999; Weber et al. 2001).

Some authors, such as Kalvin et al. (1995) and Zollikofer (2002), state the importance of the Computer-Assisted Anthropology/Paleoanthropology (CAA/CAP) devised as a combination of noninvasive methodologies to reconstruct fragmentary fossil specimens electronically, in the spirit of reverse engineering. CAP combines medical imaging with computer graphics and rapid prototyping technologies: three-dimensional data is acquired with Computer Tomography (CT) which permits non-invasive sampling of external and internal data from the fossil specimens.

Computer-assisted acquisition of data from fossil specimens has attained a high level of efficacy, accuracy, and reproducibility, and provide tools to display, manipulate and measure fossil specimens on screen.

Kalvin et al. (1995) focus on several important advantages of the CAA over the physical reconstruction method:

- Plan and execute reconstructions more accurately
- Create and modify existing reconstructions with greater ease and flexibility
- Perform more sophisticated quantitative analysis of fossil specimens
- Make better use of the relatively small number of fossil pieces that have been discovered

The use of CT, helps in the reconstruction of incomplete fossils by using mirror-imaging, assist in the extraction of embedded rock matrix fragments non-destructively, using image processing to segment out the matrix (Kalvin et al. 1995; Weber et al. 2001). This greatly increases the opportunities for making composite reconstructions as it is possible to artificially segment distorted portions of any specimen, and to place each portion into a more anatomically parsimonious position, thereby partially removing plastic distortion (Neeser 2007). Further in the chapter, we can find examples for these cases, applied to cranial studies.

Application of the Computed Tomography on Human Fossil Crania

Phillip Tobias (2001), displays the beginning of the application of the computed tomography on human fossil crania, from the introduction of CT in the 1970s, and how two teams, one in The Netherlands [lead by Wind and Zonneveld (1985,

1989)] and one in United States [Conroy and Vannier (1984, 1987, 1991)], developed this method along the 1980s decade. In July of 1990 these two teams worked side by side in Johannesburg (Franz W. Zonneveld pers. comm.).

Since then, CT was considered a very useful tool to study fossil hominids skulls, and the internal structures that were inaccessible to other techniques, as was demonstrated at the end of the 1980s and along the 1990s in several works (Conroy et al. 1990, 1998a, b; Zonneveld et al. 1989; Conroy and Vannier 1987; Zollikofer et al. 1995).

Some years before, other authors showed firsts attempts using CT. It is advisable here to mention works as Legoux (1966), Delattre et al. (1967), Price and Molleson (1974) or Hotton et al. (1976).

CT is therefore a helpful tool for studying intracranial surfaces as a nondestructive manner, so we will be able to register the internal structures in sufficient detail to reconstruct later (Zonneveld and Wind 1985).

The visualization of previously hidden intracranial morphology, and the precise geometric data obtained with the CT scanner, extended the possibilities in the study of Human fossil crania.

Wind and Zonneveld (1989) performed the first study of an *Australopithecus* Skull "Sts 5" (Mrs. Ples). They describe the use of an advanced X-ray method, the computed tomography, normally used in the 1980s decade for clinical diagnosis.

This technique, allowed not only the knowledge of the internal morphology of the fossil hominid skull, but the reconstruction of various functional characteristics of the hominid (described in the Introduction).

CT demonstrated the intra-osseous morphology of the *Australopithecus* skull, that had only been described externally. The high quality of the CT images, allowed the visualization of small density differences thus resulting in highly detailed cross-sectional images.

The following works, represent the earliest attempts for the development of some techniques which are now very advanced. That is the case of the australopithecine partial cranium MLD 37/38.

The first digital study of this specimen, required the development of a new technique in order to restore missing portions of the braincase and endocast.

A number of contiguous transverse CT slices, were separately treated, tracing the contour of the interface between the matrix and the endocranial surface of the calvarial bone. The software calculated the area in cm^2 of the established endocranial region in each slice. When this value was multiplied by 0.2 cm (the distance between the slices), the endocranial slice volume in cubic centimeters was obtained. Missing portions of the cerebrum in some slices, were restored by manual drawing. The volumes of the slices for all of the slices were stored. The sum of the slices' volumes gave the total endocranial volume of the specimen MLD 37/38. A new and accurate total volume value of 425 cm^3 was obtained (see Conroy et al. 1990; Tobias 2001 for more details). The use of CT, has allowed the analysis of the cranial venous outflow patterns, and the results support the view that gracile and robust australopithecines evolved differently in response to erect postures.

In a subsequent reconstruction of the same specimen, the fossilized bone was likewise mapped as distinct from stone matrix, for every slice of the computed tomographic image, creating a virtual endocast. In this case, missing endocranial regions, were reconstructed using thin-plate spline warping, from a complete reference specimen (see the forward point "Virtual cranial reconstruction" for more details), STS 5 (Neubauer et al. 2004).

They used a large number of landmarks, and obtained an estimated volume for the reconstructed braincase of 440 cm^3. This study concludes with the establishment of the endocranial capacity of MLD 37/38, within the range of other *Australopithecus africanus* specimens (428–515 cm^3), "had similar dimensions to STS 5 and the absence of an enlarged occipital-marginal sinus system, that is typical of *A. africanus*.

Another *A. africanus* specimen, Stw 505, was reconstructed using the mid-sagittal plane, and an accurate endocranial volume was calculated as 513 cm^3. The same method used in MLD 37/38, (calculation of the volume of each CT slice) was applied to Stw 505. The endocranial volume was 518 cc in this case (Tobias 2001).

A former 3D visualization of hidden structures in hominid fossil skulls using CT, can be observed in the Chap. 12 of the book titled "The paranasal sinuses of higher primates: development, function, and evolution" and dedicated to *Computed Tomography-based Three-Dimensional imaging of Hominid Fossils: Features of the Broken Hill 1, Wadjak 1, and SK 47 Crania* (Spoor and Zonneveld 1999).

Detailed features are displayed forward in the text.

But one of the earliest applications of CT to the study of internal features of hominids skulls, is represented by the work of Zonneveld and Wind (1985) and Zonneveld et al. (1989). In these works they exhibit the best parameters of use achieved at that time, describe the method, the technique, explain problems with visualization of matrix filled structures, and symbolize a first step for the improvement of this type of study. Furthermore, a high number of fossil specimens are included in their analysis.

The next points, correspond to some detailed studies of different cranial structures, as an example of the large potential area of study for the method.

Dental Analysis

One important topic in the study of human evolution, is to test and compare the development and growth rate in hominoids species. To examine the state of maturation of an individual, a dental analysis is necessary, in order to make studies of palaeodemography and social behaviour based on dental eruption patterns. The enamel thickness patterning provides valuable taxonomic, functional, and/or phylogenetic information and provides unique information regarding the timing of dental maturation that can be used to compare and contrast developmental patterns between extant and extinct hominoids (Conroy and Vannier 1991; Schwartz 2000; Olejniczak et al. 2008). Furthermore, dental enamel provides information about diet and other behavioural patterns.

First applications of CT to the study of cortical bone and dental enamel, tells about the accuracy of the measurements, but makes reference to the problems associated with the highly mineralized fossils (Spoor et al. 1993).

The problem of CT-scale overflow can also be caused by the scanning of small objects (the lack of beam hardening in the small object creates CT numbers that are much higher than what can be expected in patients).

For this reason we always scan small objects placed within a hollow cylinder that surrounds the object with a total of 10 cm plexiglass, which takes care of the required additional beam hardening (Franz W. Zonneveld pers. comm.).

Conroy and Vannier (1987) investigated the Taung child's skull maturation, using computerized tomography, concluding that this specimen shows some

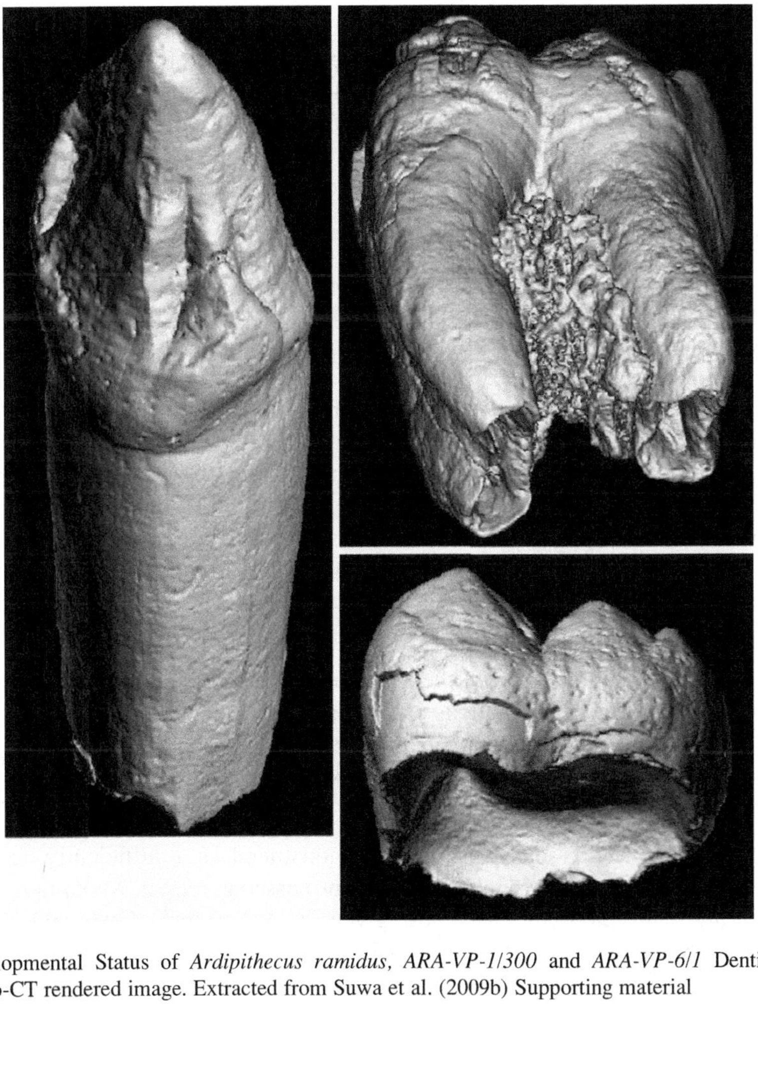

Developmental Status of *Ardipithecus ramidus, ARA-VP-1/300* and *ARA-VP-6/1* Dentitions. Micro-CT rendered image. Extracted from Suwa et al. (2009b) Supporting material

important dental maturational affinities with great apes, although, as Dart noted, other hominid-like features are clearly present. One of the latest studies applying 3D Micro-CT, focused on the remains of *Ardipithecus ramidus*, (Suwa et al. 2009b) and concluded a different diet in this species from that of chimpanzee with lack of thick enamel, and dentition not adapted to abrasive food.

Tooth enamel thickness has long been an important character in studies of primate and especially hominin phylogeny, taxonomy, and adaptation. Macho and Thackeray (1992) highlighted differences in enamel thickness over functionally significant regions of the crown in a group of hominids. Differences in the distribution of enamel in *A. robustus*, *A. africanus*, and *Homo* sp. were identified and interpreted in terms of dietary regimes.

Microfocal X-Ray Computed Tomography, is an accurate technique for measuring enamel thickness in recent taxa, although heavily mineralized teeth pose an obstacle to the ability of micro-CT to distinguish dental tissues. Moreover, extremely thin enamel (less than 0.10 mm) is difficult to resolve adequately in raw micro-CT images based on pixel values alone. Therefore, caution must be exercised in the application of micro-CT to the study of fossilized teeth (Olejniczak and Grine 2005, 2006; Olejniczak et al. 2007a, b).

Bony Labyrinth of the Inner Ear

The inner ear is a structure located inside the petrous temporal bone, and it is also known as labyrinth because of the complexity of its shape. This structure houses the organs of hearing and balance (Spoor and Zonneveld 1998). It consist of two parts, the osseous labyrinth and the membranous labyrinth. The osseous labyrinth in turn, consist of three parts; the vestibule, semicircular canals and cochlea. The membranous labyrinth is contained within the bony part (Gray 1995).

The bony labyrinth, leaves an empty space in the temporal bone that can be filled to obtain a cast for subsequent studies.

CT scans allow three basic points of study, earlier considered on physical casts (Spoor and Zonneveld 1995), and enhance the measuring procedure and the results:

1. Descriptive and comparative works
2. Biophysical relationship with the function of the vestibular organ, the dimensions and the planar orientations of the semicircular canals
3. Angular and distance measurements to describe morphological features of the labyrinth that are said to be related to the ontogenetic and phylogenetic development of the cranial base

The use of the CT scans, has been demonstrated as a sufficiently accurate method for morphometric analysis of this intraosseous region. Maximum errors comparing with physical casts, are given in Spoor and Zonneveld (1995) with very low values as $\pm$ 0.1 mm for linear dimensions, $\pm$ 4° for orientations and $\pm$ 2% for indices describing size and shape of the bony labyrinth. This is an

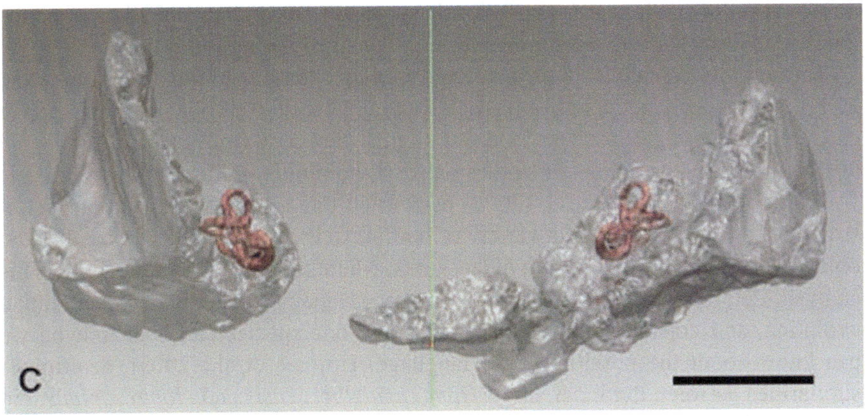

Inner ear of *Ardipithecus ramidus*, ARA-VP-1/500. Extracted from Suwa et al. (2009a)

organ with a complex three-dimensional shape, and CT enables a fast and easy reconstruction.

Aspects of locomotor behaviour, erect posture and the kinematic properties of the head and neck are some of the questions that the study of the early hominids bony labyrinth can answer (Spoor et al. 2003).

Previous studies using high-resolution computed tomography applied to this cranial region, result in very interesting conclusions.

An early comparative work, using a large sample of fossil hominids and apes, concluded that the inner ear is largely similar in function and morphology in humans and apes (Wind and Zonneveld 1985).

Many modern studies, clearly determines *H. erectus* as the earliest fossil hominids species to demonstrate the modern human morphology, and *Australopithecus* and *Paranthropus* resemble those of the extant great apes (Spoor et al. 1994). Neanderthals semicircular canals may reflect a pattern of head movements different from that of modern humans (Spoor et al. 2003).

Description and comparison of the Nazlet Khater 2 (NK 2) inner ear (this is the only complete modern human skeleton from the earliest Late Stone Age in Africa) by Bouchneb and Crevecoeur (2009), establishes the morphology and biometrical characteristics of the NK 2 bony labyrinth compared to extant modern humans.

The NK 2 inner ear presents particular features that are unusual within our extant modern human sample. These results suggest that Late Pleistocene modern human variation differs from that of recent humans.

Other no functional conclusions can be considered too. As CT reveals more details than traditional X-ray procedures, Wind (1984) concluded, from an anatomical study of the internal structures (temporal bones) of a rather heavily mineralized *H. erectus* skull, i.e., *Pithecanthropus IV*, that the analysis of the right temporal bone shows an extensive pneumatization. Specifically, that the oval shape of the external meatus does not result from fractures as previously suggested, and that the middle ear and the internal auditory meatus have probably been damaged during restoration.

The auditory capacities of primates and specially some hominids (Martinez et al. 2004), have been considered to answer some doubts on the topic of the language capacities in these groups. This idea was slightly proposed by Wind and Zonneveld (1985) in the first radiological studies on fossil hominids skulls.

In this way, Martinez et al. (2004) performed three-dimensional virtual reconstructions of the inner ear of a variety of hominids, and they have been used as the object of the study, in contrast to older works, where the hyoid bone was used. This analysis determines a new approach to the language capacities in the human evolution, and considers that, if we know how one species can hear, maybe we can know about the possibilities of language. Urquiza et al. (2005), determines similarities between the inner ear of *Homo heidelbergensis* and *Homo sapiens*, but Martinez et al. (2004) focuses on the conclusions, and shows that the skeletal anatomy of *H. heidelbergensis* is compatible with a human-like pattern of sound power transmission through the outer and middle ear at frequencies up to 5 kHz. This suggests that they already had auditory capacities similar to those of living humans in this frequency range.

Thompson and Illerhaus (1998), used 3D-micro-CT as it provides higher resolution than other CT methods for the study of the bony labyrinth of the inner ear of Le Moustier 1 Neandertal adolescent. The resulting geometry allowed for a much more accurate measurement of these internal structures of the skull, that could be manipulated in 3D space in order to measure the dimensions of the radii of curvature of the semicircular canals and the position of the posterior semicircular canal relative to the plane of the lateral canal following published methods (Spoor and Zonneveld 1995; Hublin et al. 1996).

The following work, from Ponce De Leon and Zollikofer (1999), also analyzed the semicircular canals on both sides, concluding the exhibition of an array of morphometric features of special interest. They described the common crus of the superior and posterior canals sloped steeply, yielding a fairly low position of the lateral canal relative to the posterior canal. Moreover, the lateral semicircular canal slanted downward toward the rear.

Peculiarities and Problems Derived of the Use of CT Scans

Zonneveld and Wind (1985), found some peculiarities early in the application of CT scans for the study of the intracranial morphology. In the case of the labyrinthine structures, calcite and other minerals are deposited inside the labyrinthine cavities during the fossilization process, and as its density is higher than the bone, the structure is lighter in the CT scans, but remains a very narrow air space in the centre of the canals. They determined a change in the standard CT-numbers, [normal range from −1,000 (air) to +3,095 (human tooth enamel)] because highly mineralized fossils were difficult to interpret in terms of these values. A new scale, ranging between −1,000 and +7,191 could solve the problem.

In addition, it is useful to mention that one of the difficulties with tomography, was that when dense matrix existed outside the imaging plane, this would cast a severe shadow onto the image which drastically reduced the contrast in the image (Franz W. Zonneveld pers. comm.).

Virtual Cranial Reconstruction: Computer-Assisted Anthropology

Unfortunately, fossils often appear partly distorted from fracturing, displacement and plastic deformation, due to taphonomical processes.

Other fossil pieces, cannot be physically extracted from rock matrix in which they are embedded without seriously damaging them (Kalvin et al. 1995; Zollikofer 2002). This is involved in the incompleteness and fragmentary state of some individuals, and result in limited analysis and scarce data from the fossils.

Reconstruction is the first step to start any comparative analysis (Gunz et al. 2009). To solve fragmentation and distortion, results in more data and a correct morphology of the specimens to arrive at new conclusions, and to locate species in a right phylogeny. But physical reconstructions, using a variety of materials and procedures to stabilize parts and complete the missing regions, usually damage the fossil (Ponce De Leon and Zollikofer 1999).

Computed tomography, together with appropriate software, can be used to virtually reconstruct skulls from various fragments without damaging the original specimen, and can extract embedded fragments non-destructively, using image-processing to segment out the matrix (Kalvin et al. 1995; Thompson and Illerhaus 1998; Ponce de León 2002; Zollikofer et al. 1998).

Once the virtual reconstruction is carried out, then the deformation can be considered and solved. The fragments, previously scanned and virtually reconstructed, are carefully oriented and located in the virtual space, following the morphology and other anatomical features in the bones.

In the case of incomplete and deformed specimens, the virtual reconstitution can be executed using two models (a) the own specimen or (b) a reference specimen/sample. Gunz et al. (2009) explored all the possibilities for cranial reconstruction. We just briefly describe some of the points they consider; for further data see the referenced paper.

The Use of the Own Specimen

As Gunz et al. (2009) explain, the crania are practically symmetric. The more asymmetry there is in the specimen, the greater confirmation of the existence of deformation. For the correction of this deformation, at least one intact side and the midline are needed. Bilateral symmetry will be restored using the mirror-image procedure, which is used to complete the incomplete specimens too.

First step is the localization of the midline. A mirror image of the unaffected side is placed as a reflection from the midline on the deformed side, so the bones in the distorted part can be correctly located as well, following some homologous points in both the original and reflection.

Some morphological points along the cranial bones are needed to align the pieces correctly for an accurate articulation. In the case of Le Moustier 1, Ponce de León (2002) used the suprainiac fossa and the cranial contour for a correct articulation of the cranial vault bones, and a small portion of the jugular fossa for the correct anatomical position of the basicranial fragments.

This type of reconstruction has been applied to brain endocasts as well. Mixing virtual and physical reconstruction, the brain endocast of Taung child was completed, and new measurements and the volume were calculated (see Falk and Clarke 2007).

Some Famous Cases

Le Moustier 1 specimen, is a Neandertal male adolescent with a long history, and at least 4 previous physical reconstructions, from 1908 to the early 1920s (see Thompson and Illerhaus 1998; Ponce de Leon and Zollikofer 1999; Ponce de León 2002).

This is a very complete Neanderthal specimen so it is also very interesting for studies about ontogeny. The distortion in the cranial base (plastic deformation and fractures), is responsible for the exhibited asymmetry in the reconstructions of the cranial vault. Ponce de León (2002) corrected this asymmetry by positioning the virtual skull in the original orientation in which the actual skull was discovered, and extending the virtual skull in the vertical direction reaching the symmetry with respect to its midsagittal plane. Then, she re-established the anatomical contacts between the isolated fragments (see more details in Ponce de Leon and Zollikofer 1999).

Along this work, she detected not only the taphonomic deformation, but in vivo modification too.

Stw 505, is a partial cranium of *A. africanus*. The cranium displays some fracture-distortion in the vault. On the assumption that the vault was symmetrical, an undistorted three-dimensional model of the cranium was generated, with the determination of the mid-sagittal plane.

In this way, the volume of the virtual endocast was calculated too (see Conroy et al. 1998b; Tobias 2001). This case is explained in the point "*Application Of The Computed Tomography On Human Fossil Crania*".

The Use of a Reference Specimen/Sample

Defined as Geometric reconstruction in Gunz et al. (2009), it is based in geometric morphometric analysis, using the properties of the Thin-plate-splines. It requires curves and surfaces established from a reference set of data, and will be deformed by the target specimen.

To estimate the missing coordinates on the target form, a thin-plate-spline interpolation is computed from the subset of the landmarks and semilandmarks available in both the complete reference and the incomplete target specimen. This interpolation function is used to map the missing landmarks from the reference onto the target.

The statistical reconstruction is considerable as well, because *predicts the location of every missing coordinate using multiple multivariate regressions based on a sample of complete specimens*, Gunz et al. (2009).

This method was used in the case of the partial cranium MLD 37/38 study (previously considered at the beginning of this section), using the reference specimen STS 5, *A. africanus* (Tobias 2001; Neubauer et al. 2004). Both specimens were virtually represented.

> The digital reconstruction used here relies on a deformation of the reference specimen's endocranial surface by thin-plate spline warping. The endocranial surfaces were represented by about 100 000 points on each surface. The endocranial surface of STS 5 was warped to that of MLD 37/38 using the homologous landmarks and semi- landmarks. This resulted in an exact match of the semi-landmarks. The missing frontal part of the endocast of MLD 37/38 was then estimated using the thin-plate spline interpolation of the available part, which deformed the STS 5 endocranium accordingly (Neubauer et al. 2004).

Zollikofer et al. (2005) made the reconstruction of TM-266-01-60-1 specimen, attributed to *Sahelanthropus tchadensis* species, using both previously described protocols for the restoration:

Protocol A *used features shared by all mammal crania to position and orient each fragment. First, the basioccipital was positioned and oriented in the midsagittal plane. The temporals were then adjoined from both sides and aligned by placing all of the left and right semicircular canals in approximately parallel orientation. Lateral and superior parts of the vault were adjoined by using the well-preserved temporal lines to establish bilateral symmetry. Within the face, displaced but undistorted portions of the supraorbital torus and orbital margins were repositioned symmetrically relative to the midsagittal plane. Left–right asymmetry in the maxilla from plastic taphonomic deformation was partly corrected with the use of published methods.*

Protocol B *used a geometric approach based on stepwise reduction of degrees of freedom of the position and orientation of individual parts relative to each other. This method takes advantage of the almost complete preservation of the TM 266 cranium, in which the position and orientation of each fragment is spatially constrained by contacts with all neighbouring fragments, and overall morphology is constrained by bilateral symmetry. Translational degrees of freedom were first reduced by re-establishing morphological continuity between dislocated fragments along matching fracture lines (along the nuchal plane, along cracks in the right parietal, between parts of the supraorbital torus, and between dislocated parts of the midface). Rotational degrees of freedom between adjacent fragments were then reduced by stepwise integration of fragments into the reconstruction, followed by iterative adjustments until a symmetrical integrated morphology was achieved.*

These procedures were applied to orient left and right neurobasicranial sides relative to each other, and the maxillae relative to the midface.

Last, deviations from bilateral symmetry in the maxilla were partly corrected as in protocol A.

In both protocols, the face and the braincase were reconstructed independently and then assembled by using anatomical continuities within the squamous portions of the frontal bone; along preserved continuities between the basisphenoid, the pterygoid processes and maxillary tuberosity on the right side; and between the bones of the right temporal fossa (squamous sphenoid, zygomatic, maxilla and frontal).

Endocranium

Brain Endocast

The brain, during its development, leaves marks on the internal cranial surface such as gyri, sulci, blood vessels and sinus, because of the tight structural relationship between them during the neurocranial growth process.

This correspondence between the inner table and brain makes the endocranial cavity a very useful cast of the cerebral surface and volume, as well as of the supporting structures.

The endocraneal surface is then a useful record of the pre-existent soft tissues (Bruner 2003), so we can create a reference of how the brain was in vivo, then it is possible to get additional information about the exocranial surface in comparison to the traditional study.

The brain, during normal growth, pushes the bones of the skull, so they take the shape established by the brain, from tensions created by the attachment fibers of the dura septa, which are closely related to the sutural system of the cranial vault (Hoyte 1997).

However, the external resemblance of skulls may not always correlate with endocranial similarity (Seidler et al. 1997).

That is why the study of the endocast is very interesting, because they complete the analysis provided by the skulls.

In some cases, taphonomic processes undergone by skulls, have resulted in natural endocasts, because the endocranial cavity has been filled with the fine soil matrix where it has been embedded, and the sediment solidifies and fossilizes (Falk 1987), as in the case of the famous Taung Child (*A. africanus*), published by Raimond Dart in 1925, MLD 37/38 (*A. africanus*) also published by R. Dart in 1959 (Neubauer et al. 2004), Sts 60 or Sterkfontein n° 1 (*A. africanus*) by Broom and Schepers in 1946 published together to Sterkfontein n° 2 and n° 3, Sts 58 (without Sts 19 with which it was related in 1950 by Broom y Schepers) described by Dean Falk in 1980, Sts 1017 by D. Falk in 1979 or SK 1585 by Holloway in 1972 (Falk 1980).

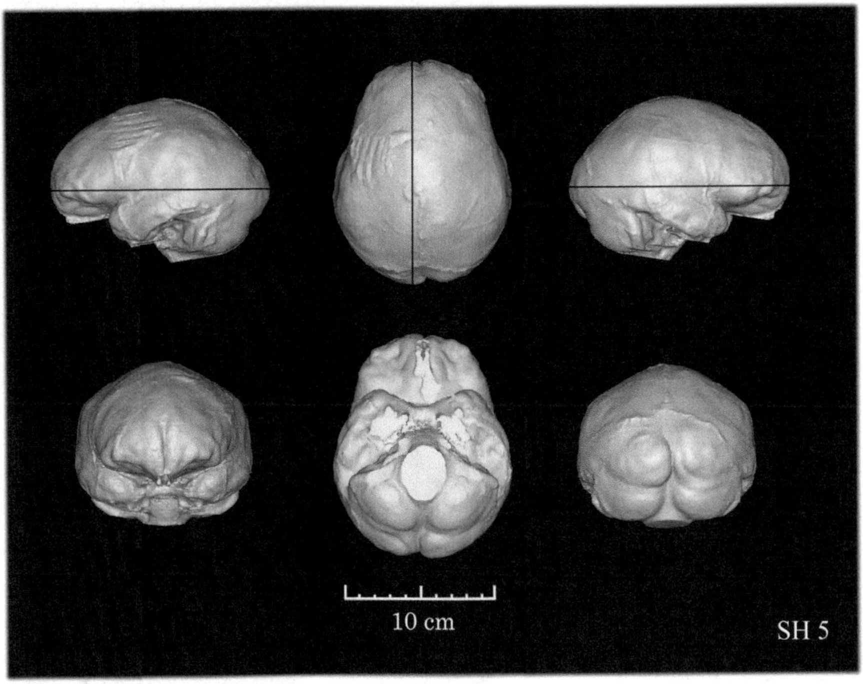

Brain endocast of Cranium 5 from Sima de los Huesos (Atapuerca)

But most crania are too fragmented or damaged to allow for the preservation, integrity, or contiguity of many internal features (Laitman 2004). Unfortunately, the process that results in the formation of an endocast, is infrequent, and if they have come to the point of being created, they have not been found or have been destroyed by natural processes.

So, from the beginning of the study of endocasts in nineteenth century, reconstructions have been created physically, using material such as latex or plaster (Symington 1916; Keith 1931; Kappers 1936; Falk 1987; Holloway et al. 2004. . .etc.).

But this method causes damage to the fossil, because of the chemical components of the products used, and the manipulation that the fossil is subjected to.

Application of CT to the study of endocranial cavities, has yielded positive results because, the integrity of the fossils is protected on the one hand, and you can get three-dimensional virtual copies of the endocast, isolated and high quality and resolution on the other hand, allowing for numerous studies.

Combined with digital morphology and geometric morphometrics (e.g., Zollikofer et al. 1998; Recheis et al. 1999; Spoor et al. 2000a, b), it allows for enhanced access to several unexplored volumes and surfaces (Bruner and Manzi 2005).

Since they are precise, parameters such as linear distances, angles, areas, and volumes can be accurately (and instantaneously) assessed.

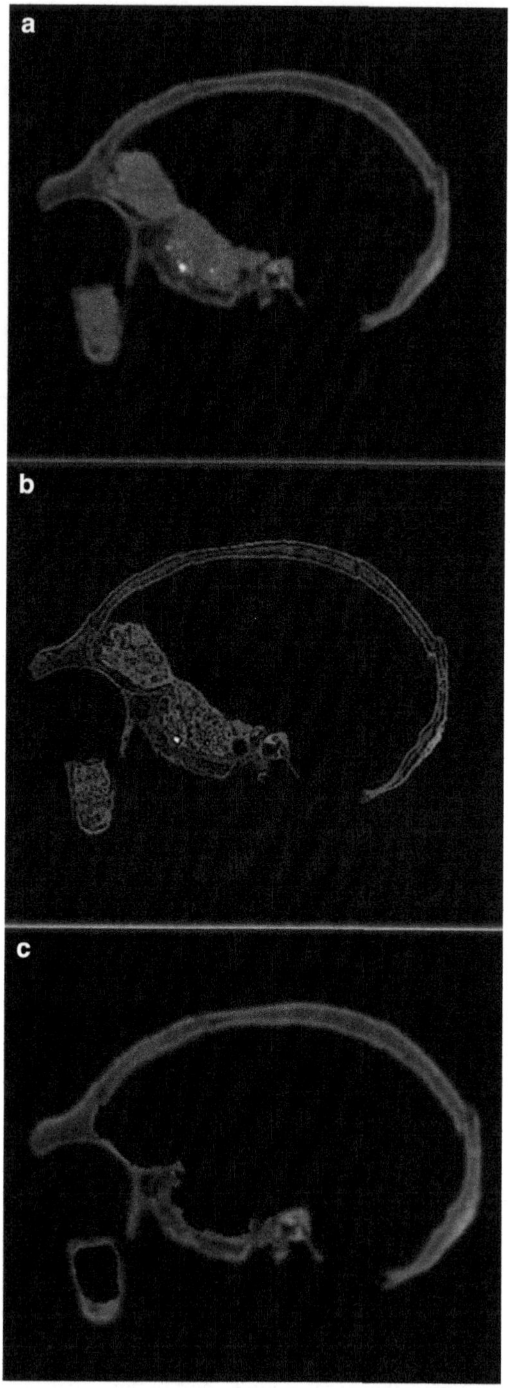

Segmentation process: the sequence of steps that lead to a segmentation of the fossil cranium from its encrustations. (**a**) One slice of a Mid-Pleistocene cranium together with the encrustations. The white extended dots are attenuation images of pebbles. (**b**) The resulting slice image after the sequence of filter steps have been applied. Note that the filter algorithm also finds a boundary between the largest pebble and the rest of the encrustation. The sediment is clearly isolated. (**c**) The result after the image-editing removal of those pixels/ voxels that were identified by the filter algorithm to be encrustation. After removal of the sediment image, the cranium reveals interesting endocranial features. Extracted from Weber et al. (2001) and Prossinger et al. (2003)

The procedure of obtaining the virtual endocast, requires an appropriate software (frequently used in Medicine and Engineering), such as MIMICS, VOXEL-MAN, AMIRA, which, besides the tomographic image visualization, allows for obtaining the volumes of the several intracranial cavities, that can be filled with matrix or air.

The difference between the density values (Hounsfield units, HU) of the bone, and the air or matrix, makes it easy to isolate the corresponding cerebral cavity area in each of the tomographic slices. Manual editing is used to isolate the empty space (air) of the cerebral cavity from the surrounding bone. This process is named segmentation. The selection of the values in the HU threshold scale, will result in a higher/lower final resolution of the virtual object, so a right selection will result in a better detailed brain endocast. A color and opacity is assigned to the selected voxels.

The last step is 3D rendering, then the software automatically links the values selected in all the slices, and results in a 3D object with total volume of the endocranial cast. CT allows volume rendering, that is the most complete volume data set. At the end, we obtain an exact copy of the endocranial details.

Once the virtual brain endocast is performed, we can analyze morphology, size, volume and middle meningeal vessels, as part of human evolutionary studies.

In this way, CAP has led the analysis of the endocranial volumes and brain endocast surfaces and, together with the use of software designed for 3D CT reconstruction, the study of internal features and the generation of virtual brain endocast of several skulls in the fossil record have been possible, such as some Dmanisi specimens, Sambungmacan 3, Bodo, Petralona, Broken Hill 1, Saccopastore, Le Moustier 1, Mojokerto, Hexian, LB1, the last virtually reconstructed Taung child, SH5 (Atapuerca), Zhoukoudian sample…etc. (Grimaud-Herve et al. 2006; Broadfield et al. 2001; Conroy et al. 2000; Seidler et al. 1997; Falk et al. 2005; Manzi et al. 2001; Bruner et al. 2001; Ponce de Leon and Zollikofer 1999; Balzeau et al. 2005; Wu et al. 2006; Falk and Clarke 2007; Poza-Rey and Arsuaga 2009; Wu et al. 2010).

Middle meningeal vessel imprints, have led some paleoneurologists, such as Emiliano Bruner, to focus on the morphology of these structures, leading to inferences on the cerebral physiology and metabolism in extinct human groups (Bruner et al. 2005). He hypothesised on the evolution of these structures as variation in the middle branch which supplies the parietal areas. Others confirm previous conclusions as dominance of the posterior branch over the anterior one in some extinct fossil groups (Grimaud Hervé 1997) and a definite increase in the number and complexity of the anastomoses and reticulation in modern humans (Bruner and Sherkat 2008) already observed in physical endocasts (Saban 1984). In the last revision to this works, authors tested the relationship between neurocranial shape and the general morphology of the traces of the middle meningeal vessels in a modern human population, by using landmark-based geometrical models. The conclusions established important points to be considered: *Although there are some neurocranial differences between groups with different vascular patterns, they are very small or not statistically significant. Even if the neurocranial differences among extinct hominids are definitely larger than those within the modern species, the present analysis suggests that it is unlikely that the differences in vascular patterns among the human species are related only to the effects of different neurocranial geometry* (Bruner et al. 2009).

Endocranial Features

Other internal features of the cranium, are also available with virtual reconstruction. Some specimens, such as crania filled with heavily calcified matrix, could not be studied in the past. *Ordinary radiographs of such specimens usually failed to reveal endocranial structure, as fossilized bone and calcified endocast were approximately equally radio-opaque.*

Thus, neither endocranial volume nor structural details were detectable (Tobias 2001). Nowadays, CT has been successfully applied to paleoanthropology, and matrix is virtually deleted allowing the access to internal traits.

In this case, selected voxels for the virtual cranium reconstruction correspond to the density values (HU) of the bone. This is an easier and faster process than the case of the virtual brain endocast, because processing all the slices is unnecessary. A virtual cranium, is very easy to manipulate, so you can, e.g., virtually open the calvaria and study the internal structure of the cranial base. The excellent work that focused on Eliye Spring specimen (Bräuer et al. 2004), represents a good example of the endocranial analysis possibilities, with a comparative study of some fossils and modern human crania. Eliye Springs is usually associated with "archaic" *H. sapiens*. This group of ancestors are particularly interesting because, while they represent early members of our own species, they also appear to retain ancestral features.

Clearly, the mosaic of modern/premodern features that complicate the external surface of "archaic" *H. sapiens* extends to key internal landmarks as well (Laitman 2004).

Weber et al. (2000), made an attempt to apply CT as a new approach for the analysis of cranial bone thickness, because information about the thickness of cranial bones is not only of great medical interest, but can be just as useful for investigations of fossil hominid material (Gauld 1996). Previous application of CT to cortical bone was already tested in Spoor et al. (1993), and useful in the study of frontal cranial profiles in Bookstein et al. (1999).

Pneumatization of the Paranasal Sinuses

Pneumatization by definition is the presence or development of air filled cavities in a bone (Al-Faleh and Ekram 2005). The major or minor development of these cavities, may in some cases explain the external appearance of the cranial structures.

Many studies have claimed specific morphological conditions, such as the massive pneumatization of sinuses, as significant for the phylogenetic interpretation of particular hominids (Bräuer et al. 2004).

Former studies about the facial sinuses are scarce; few skulls available showed these areas of interest (Keith 1931; Weidenreich 1946). Until only a few decades ago, anthropological examinations were restricted by the limited access to these fossils.

Some specimens with highly calcified matrix were impossible to analyze. Conventional radiography was not able to reveal endocranial structure because of the resemblance of the fossilized bone and matrix gray values.

Many specimens have shown various degrees of incrustation caused by stone matrix. These incrustations are represented with (nearly) the same Hounsfield Units as the parts of interest lying beneath. Generally there were no easy methods to remove these disturbing incrustations.

Thus, neither endocranial volume nor structural details were detectable (Tobias 2001).

CT imaging made it feasible for the first time to analyze and obtain images of the endocranium, sinus cavities and inner ear embedded in soil matrix, which had not been identified on X-ray films (Mafart et al. 2004). CT image acquisition and 3D-imaging development, brought progress to the study of the facial sinuses.

Sophisticated filters have being designed to get access to the underlying parts of interest. The skulls of Steinheim as well as Bodo and OH 9 are prominent examples. Newly designed algorithms were used to remove the incrustations of the frontal sinuses (see Recheis et al. 1999).

The case of the Eliye Springs cranium required as well virtual removal of the sandstone matrix from some internal cavities. Since the fossil bone and the matrix exhibit close similarities in density and fall into the same intensity range, the different objects often had to be separated by manual editing on orthogonal (transverse, sagittal, and coronal) CT slices. Difficulties in separating bone from matrix occurred in this case especially with regard to the sphenoidal sinuses (Bräuer et al. 2004).

Frontal Sinus

Results of studies of the frontal sinus, have been applied to ontogeny, as in the case of Le Moustier 1 (Thompson and Illerhaus 1998), revealing that the morphology in this juvenile specimen is similar to that found in adult Neanderthals. Frontal sinus was once more considered as a criterion in estimating the *probable range* of age at death for Krapina 1 (Minugh-Purvis et al. 2000). It would appear, based on radiographic examination, that the frontal sinus development in Krapina 1 was comparable to that found in the La Quina 18 and Teshik-Tash 1 juveniles, aged at 7.5 and 10 years, respectively (Minugh-Purvis et al. 2000).

ATD6-15, is afrontal bone of *Homo antecessor* from Gran Dolina in Atapuerca site (Bermúdez de Castro et al. 1997; Arsuaga et al. 1999). The frontal sinus of this specimen, are fairly extensive, but this individual has been considered to be a subadult, because of other cranial features, and by comparison with the material of Sima de los Huesos in Atapuerca site (Arsuaga et al. 1997) and

Neandertal (Smith and Ranyard 1980) samples of immature frontal bones. In modern humans, the main enlargement of the frontal sinus is completed at around 16 years for boys and 14 years for girls (Arsuaga et al. 1999). Moreover, in modern humans, frontal sinus development correlates closely with supraciliary arch appearance (Weinmann and Sicher 1955), which, in turn, begins its development at approximately the time of first molar eruption (Minugh-Purvis 1988 in Minugh-Purvis et al. 2000).

A possible correlation of ATD6-15 with the pre-adolescent individual ATD6-69, suggests the frontal sinus developed at an earlier age in *H. antecessor* than in modern humans (Arsuaga et al. 1999). Despite considerable variation in shape, pneumatization and maturational course, frontal sinus development in modern humans is minimal until mid-childhood (Szilvássy unpublished data, 1981 in Minugh-Purvis et al. 2000). This also suggests that supraorbital torus shape of ATD6-15 was still far from the adult condition, and that thickness and projection of the torus, as well as frontal squama thickness, would substantially have increased with age (Arsuaga et al. 1999).

Visualization of the internal morphology of the supraorbital torus appears similar in some ancient fossils, but reconstructions performed at the University Clinic of Innsbruck showed differences (Recheis et al. 1999).

Three Middle Pleistocene specimens, Broken Hill 1 (Kabwe), Petralona and Arago 21, were compared in Seidler et al. (1997). The analysis of the paranasal sinuses, resulted in several conclusions. Broken Hill 1 (Kabwe) and Petralona have extremely large frontal sinuses. Postorbital constriction is much more accentuated in Arago 21 and Broken Hill 1 (Kabwe) than in Petralona. It appears that the enormous pneumatization of the lateral frontal sinus of Petralona have decreased postorbital constriction by expanding the lateral walls of the frontal bones outward.

Significantly, in Petralona and Broken Hill 1 (Kabwe), the roof of the orbits is formed by the bottom of the frontal sinuses rather than the orbital plates of the frontal bone, as is the case for Arago 21 (in spite of its well-developed supraorbital torus) and modern humans (Bräuer et al. 2004). In other words, the frontal lobes of Petralona and Broken Hill 1 (Kabwe) are not expanded rostrally over the orbits in conjunction with an expanded forehead (Seidler et al. 1997; Lieberman 1995).

The forehead in these two specimens (*H. heidelbergensis*) is built mainly by the extent of the lower part of the frontal sinus. The forehead of modern humans is built by the anteriorly positioned frontal lobes, directly above the orbits (Recheis et al. 1999; Bräuer et al. 2004). In Eliye Springs specimen, in which the supraorbital torus is mostly broken off, it is clearly evident that the frontal lobes are placed above the orbital cavities, very similar to the placement in modern humans (Bräuer et al. 2004).

But the frontal sinus in Ceprano specimen are scarcely developed compared to the Middle Pleistocene range of variation described in Seidler et al. (1997) and Prossinger et al. (2003) and shows a small frontal sinus compared also to those in modern humans (Weiglein 1999).

It may be hypothesized that they cannot extend backward and upward, growing through the frontal bone, because of the structural independence between the supraorbital torus and the frontal squama (Bruner and Manzi 2005).

Frontal sinus development cannot be used to estimate phylogenetic affinities, at least in European human evolution (Arsuaga et al. 1991). Size and shape of the frontal sinuses show great variation in recent and fossil humans (Spoor and Zonneveld 1999; Seidler et al. 1997). Eliye Springs exhibits well-developed frontal sinus, which might have been restricted to the supraorbital torus region not invading into the squama in contrast to Broken Hill 1 (Kabwe) or Petralona (Bräuer et al. 2004).

Maxillary Sinuses

Previous hypotheses of maxillary sinuses size evolution have proposed one or more changes in the volume of the structure across hominoid phylogeny (Rae and Koppe 2000). Findings show that maxillary sinuses size correlates well with craniofacial size in all primates and in humans (Rae and Koppe 2002).

Eliye Springs hominid deviates from modern anatomy in a number of features, which include heavily pneumatized maxillary sinuses (Bräuer et al. 2004).

A buttressed canal that conducts the nerve supply for the upper front teeth may be interpreted as a secondary adaptation, or a consequence related to enlarged maxillary sinuses in Petralona and Broken Hill 1 (Kabwe) (Seidler et al. 1997).

Sphenoidal Sinuses

Sphenoidal sinuses are large in modern humans. But the sphenoidal sinuses of Petralona, are extremely large (Seidler et al. 1997). Indeed, pneumatization is so extensive in Petralona that it extends beyond the sphenoid to other bones such as the apex of the petrous bone, anterior part of the basilar portion of the occipital bone, and even the temporal squama. This morphology surpasses the known range of variations in modern humans (Hajek 1926 in Seidler et al. 1997).

Pneumatization in the stereolithographic model of Broken Hill 1 (Kabwe) appears to be similar to that of Petralona, but not so extreme.

As in the frontal sinus analysis, the sphenoidal sinuses are used in ontogeny. In modern humans the hollowing out of the sphenoid bone by the sphenoidal sinuses has begun by 6 years of age (Aiello and Dean 1990).

In the approximation of Cranium 6 of a 14-year-old specimen from Sima de los Huesos, the base of the pterygoid processes is not hollowed out by the sinuses, but in the adult specimens Cranium 4 and Cranium 5 the sphenoidal sinuses fill all the base of the pterygoid processes (Arsuaga et al. 1999). Based on modern human standards and the Sima de los Huesos evidence, ATD6-17 was assigned to a late adolescent or an adult specimen. However, the KNM-WT-15000 specimen shows an extensive cranial pneumatization, even in the sphenoid bone

(Walker and Leakey 1993). If the Dolina hominids had a sinus growth pattern as in *Homo ergaster*, ADT6-17 could belong to a pre-adolescent individual (Arsuaga et al. 1999).

The sphenoidal sinuses in Eliye springs, are relatively large and extends on the left side into the greater wing of the sphenoid bone up to the level of the oval foramen, and posteriorly up to the superior third of the clivus. Analysis of the modern sample into the same work, revealed that in Strauss 2 specimen, the sphenoidal sinuses are also large extending laterally into the greater wing up to the same level as in Eliye Springs and posteriorly well into the upper third of the clivus. In contrast, the sphenoidal sinuses of Mumba 4 specimen, are smaller and do not extend into the wing or clivus (for more details see Bräuer et al. 2004).

The sphenoidal sinuses of Petralona and Broken Hill 1 (Kabwe) continue from the body into the greater wings. The heavy pneumatization in Petralona is also characterized by an extension of the sinus into the medial part of the clivus and the temporal squama (Seidler et al. 1997).

The Mastoid Air Cells

The mastoid air cells extend unusually far in Petralona, and are in contact with distinct pneumatic cavities in the temporal bones. This morphology does not normally exist in modern humans, but has been described for the "*Pithecanthropus IV*" (Sangiran 4) specimen of *H. erectus* (Wind 1984). Mastoid air cells are developed, but are not as extensive in Broken Hill 1 (Kabwe) (Seidler et al. 1997).

Summary

From the beginning of application of computed tomography to the analysis of fossils, results obtained in all of these topics, have increased the knowledge and the possibilities in the study of human evolution. Many researchers, besides application of the new technologies, are helping to improve the quality of methodologies, software and results.

Nowadays, biomechanical and developmental processes are better known, and fossil preservation is highly achieved.

Certainly, CT data acquisition and 3D virtual reconstruction, are really important in the present and future of Paleoanthropology.

Acknowledgements Special thanks to Professor Frans W. Zonneveld, for his detailed review of this chapter, for his valuable advice, for helping me to understand the importance of highlighting some technical details, and for his attempts to eliminate some classic mistakes that seem to persevere within this topic. Thanks to Kathleen Waldock for the English revision. And to Atapuerca Research Project, funded by Ministerio de Ciencia y Tecnología, Spanish Government CGL2006-13532-C03-02.

References

Aiello, L. & Dean, C. 1990. *An Introduction to Human Evolutionary Anatomy*. London: Academic Press.

Aiello, L., Wood, B., Key, C. and Wood. C. 1998. Laser scanning and palaeoanthropology: an example from Olduvai Gorge, Tanzania. P. 223–236. In *Primate locomotion: recent advances*. Strasser, E., Fleagle, J., Rosenberger, A. and McHenry, H. eds. Plenum Press. New York.

Al-Faleh, W., and Ekram, M. 2005. A tomographic study of air cell pneumatization of the temporal components of the TMJ in patients with temporomadibular joint disorders. *Saudi Dental Journal* 2006;18 (Special Issue)-Abstr.125

Arana-Fdez. de Moya, E., Buitrago-Vera, P., Benet-Iranzo, F., and Tobarra-Pérez, E. 2006. Tomografía computerizada: introducción a las aplicaciones dentales. *Revista del Consejo de Odontológos y Estomatólogos* (RCOE) 11:311–322.

Arsuaga, J.L., Carretero, J.M., Martínez, I., Gracia, A., 1991. Cranial remains and long bones from Atapuerca/Ibeas (Spain). *Journal of Human Evolution* 20, 191–230.

Arsuaga, J.L., Martinez, I., Gracia, A., and Lorenzo, C. 1997. The Sima de los Huesos crania (Sierra de Atapuerca, Spain). A comparative study. *Journal of Human Evolution*. 33:219–281.

Arsuaga, JL., Martinez, I., Lorenzo, C., Gracia, A., Muñoz, A., Alonso, O., and Gallego, J. 1999. The human cranial remains from Gran Dolina Lower Pleistocene site (Sierra de Atapuerca, Spain). *Journal of Human Evolution* 37:431–457.

Balzeau, A., Grimaud-Herve, D., and Jacob, T. 2005. Internal cranial features of the Mojokerto child fossil (East Java, Indonesia). *Journal of Human Evolution* 48:535–553.

Bastir, M., and Rosas. A. 2005. Hierarchical nature of morphological integration and modularity in the human posterior face. *American Journal of Physical Anthropology* 128:26–34.

Bastir, M., Rosas, A. and Kuroe, K. 2004. Petrosal orientation and mandibular ramus breadth: Evidence for an integrated petroso-mandibular developmental unit. *American Journal of Physical Anthropology* 123:340–350.

Bermudez de Castro, J.M., Arsuaga, J.L., Carbonell, E., Rosas, A., Martinez, I., Mosquera, M. 1997. A Hominid from the Lower Pleistocene of Atapuerca, Spain: Possible Ancestor to Neanderthals and Modern Humans. *Science* 276:1392–1395.

Bookstein, F., Schäfer, K., Prossinger, H., Seidler, H., Fieder, M., Stringer, C., Weber, G., Arsuaga, J.-L., Slice, D.E., Rohlf, F.J., Recheis, W., Mariam, A.J., and Marcus, L.F. 1999. Comparing frontal cranial profiles in archaic and modern *Homo* by morphometric analysis. *The Anatomical Record (New Anat.)* 257:217–224.

Boone, JM., Geraghty, EM., Seibert ,JA., and Wootton-Gorges, SL. 2003. Dose reduction in pediatric CT: a rational approach. *Radiology*; 228:352–360.

Bouchneb, L. and Crevecoeur, I. 2009. The inner ear of Nazlet Khater 2 (Upper Paleolithic, Egypt). *Journal of Human Evolution* 56(3):257–62.

Branco, W. 1906. Die Anwendung der Röntgenstrahlen in der Paläontologie. *Abhandlungen der Königl. Preuss. Akademie der Wissenschaften*. Read 5 July 1906, Printed 14 August 1906.

Bräuer, G., Groden, Ch., Gröning, F., Kroll, A., Kupczik, K., Mbua, E., Pommert, A., and Schiemann, T. 2004. Virtual Study Of The Endocranial Morphology Of The Matrix-Filled Cranium From Eliye Springs, Kenya. *The Anatomical Record Part A* 276a:113–133.

Broadfield, D., Holloway, R., Mowbray, K., Silvers, A., Yuan, M., and Marquez, S. 2001. Endocast of Sambungmacan 3 (Sm 3): a new *Homo erectus* from Indonesia. *The Anatomical Record* 262:369–79.

Bruner, E. 2003. Fossil traces of the human thought: paleoneurology and the evolution of the genus *Homo*. Rivista di Antropologia. *Journal of Anthropological Sciences* 81:29–56.

Bruner, E., and Manzi, G . 2005. CT-based description and phyletic evaluation of the archaic human calvarium from Ceprano, Italy. *The Anatomical Record Part A: Discoveries in Molecular, Cellular, and Evolutionary Biology* 285A:643–657.

Bruner, E., and Sherkat, S. 2008. The middle meningeal artery: from clinics to fossils. *Child's Nervous System* 24:1289–1298.

Bruner, E., Manzi, G., and Passarello, P. 2001. The 'Virtual' Endocast of Saccopastore 1. General Morphology and Preliminary Comparisons by Geometric Morphometrics", p. 17-24, *In* B. Mafart, et al., eds. *XIV International Congress of Prehistoric and Protohistoric Science. Three-Dimensional Imaging in Paleoanthropology and Prehistoric Archaeology*. British Archaeological Reports International Series 1049, Liège (Belgium), September 2001.

Bruner, E., Mantini, S., Perna, A., Maffei, C., and Manzi, G. 2005. Fractal dimension of the middle meningeal vessels: variation and evolution in *Homo erectus*, Neanderthals, and modern humans. *European Journal of Morphology* 42:217–224.

Bruner, E., Mantini, S., and Ripani, M. 2009. Landmark-based analysis of the morphological relationship between endocranial shape and traces of the middle meningeal vessels. *The Anatomical Record* 292:518–527.

Bush, E. 2004. *Evolution and Scaling in Mammalian Brains*. California Institute of Technology. Pasadena, California. Thesis.

Bush, E., Simons, E., and Allman, J. 2004. High-Resolution Computed Tomography Study of the Cranium of a Fossil Anthropoid Primate, *Parapithecus grangeri*: New Insights Into the Evolutionary History of Primate Sensory Systems. *The Anatomical Record Part A* 281A: 1083–1087.

Calhoun, P., Kuszyk, B., Heath, D., Carley, J., and Fishman, E. 1999. Three-dimensional Volume Rendering of Spiral CT Data: Theory and Method. *Radiographics* 19(3):745–764.

Conroy, G., and Vannier, M. 1984. Noninvasive Three-Dimensional Computer Imaging of Matrix-Filled Fossil Skulls by High-Resolution Computed Tomography. *Science,* 226 (4673):456–458.

Conroy, G., and Vannier, M. 1987. Dental development of the Taung skull from computerized tomography. *Nature*, 329:625–627.

Conroy, G., and Vannier, M. 1991. Dental development in South African australopithecines. Part II: Dental stage assessment. *American Journal of Physical Anthropology* 86:137–156.

Conroy, G., Vannier, M., and Tobias, P. 1990. Endocranial features of *Australopithecus africanus* revealed by 2- and 3-D computed tomography. *Science* 247:838–841.

Conroy, G., Kane, A., Seidler, H., Weber, G., and Tobias, P. 1998a. Endocranial capacity of Stw 505 ("Mr. Ples"), a large new hominid cranium from Sterkfontein. *American Journal of Physical Anthropology. Supp* 26:69–70.

Conroy, G., Weber, G., Seidler, H., Tobias, P., Kane, A., and Brunsden, B. 1998b. Endocranial capacity in an early hominid cranium from Sterkfontein, South Africa. *Science* 280: 1730–1731.

Conroy, G., Weber, G., Seidler, H., Recheis, W., Nedden, D., and Mariam, J. 2000. Endocranial capacity of the Bodo cranium determined from three-dimensional computed tomography. *American Journal of Physical Anthropology* Volume 113:Pages 111–118.

Dart, R. 1925. *Australopithecus africanus*: the man-ape of South Africa. *Nature* 115:195–9.

Davidson, J. M. 1916. Localization by X-rays and stereoscopy. H.K. Lewis & Co. Ltd., London.

Delattre, A., Fenart, R., Empereur-Buisson, R. 1967. Orientation vestibulaire de crânes de Néanderthaliens par méthode radiographique. *J. Sc. Médicales de Lille* 85, p 219–222.

Doyon, D., Laval-Jeantet, M., Halimi, P., Cabanis, E., and Frija, J. 1995. *Manual de Tomografía Axial Computadorizada*. Masson

Falk, D. 1980. A reanalysis of the South African Australopithecine natural endocasts. *American Journal of Physical Anthropology*. 52:525–539.

Falk, D. 1987. Hominid paleoneurology. *Annual Review of Anthropology* 16:13–30.

Falk, D., and Clarke, R. 2007. Brief communication: New reconstruction of the Taung endocast. *American Journal of Physical Anthropology* 134:529–534.

Falk, D., Hildebolt, C., Smith, K., Morwood, M., Sutikna, T., Brown, P., Jatmiko, Saptomo, E., Brunsden, B., and Prior, F. 2005. The Brain of LB1, *Homo floresiensis*. *Science*:242–245.

Funama, Y., Awai, K., Nakayama, Y., Kakei, K., Nagasue, N., Shimamura, M., Sato, N., Sultana, S., Morishita, S., and Yamashita, Y. 2005. Radiation Dose Reduction without Degradation of Low-Contrast Detectability at Abdominal Multisection CT with a Low- Tube Voltage Technique: Phantom Study. *Radiology* 237:905–910.

Gauld, S. 1996. Allometric patterns of cranial bone thickness in fossil hominids *American Journal of Physical Anthropology* , 100:411–426.

Goldman, L. 2007. Principles of CT and CT Technology. *Journal Of Nuclear Medicine Technology* 35:115–128.

Goodman, L.R. 2010. The Beatles, the Nobel Prize, and CT Scanning of the Chest. Radiologic Clinics of North America Thoracic MDCT Comes of Age 48:1–7.

Gorjanovic-Kramberger, D. 1906. *Der Diluviale Mensch von Krapina in Kroatien: ein Beitrag zur Palaeoanthropologie*. Wiesbaden: Kreidel.

Gray, H. (1995). *Gray's Anatomy*. 15 th Edition. Pickering Pick, T & Howden, R. Editors. New York, Barnes & Nobles.

Grimaud Hervé, D. 1997. *L'évolution de l'encéphale chez Homo-erectus et Homo-sapiens* - CNRS editions.

Grimaud-Herve, D., Lordkipanidze, D., de Lumley, M.A., and Gabounia, L. 2006. Etude preliminaire des endocrânes de Dmanissi: D 2280 et D 2282. *L'Anthropologie* 110:732–765.

Gunz, P., Mitteroecker, P., Neubauer, S., Weber, G., and Bookstein, F. 2009. Principles for the virtual reconstruction of hominin crania. *Journal of Human Evolution* 57:48–62.

Hajek, M. 1926. *Pathologie und Therapie der entzuendlichen Erkrankungen der Nebenhoehlen der Nase*. Wien: Deuticke.

Hemmy, D., Zonneveld, F., Lobregt, S., and Fukuta, K. 1994. A decade of clinical three-dimensional imaging: a review. Part I: Historical Development. *Investigative Radiology*; 29 (4):489–496.

Holloway, R., Broadfield, D., Yuan, M., Schwartz, J., and Tattersall, I. 2004. *The Human Fossil Record, Volume 3, Brain Endocasts– The Paleoneurological Evidence*. Ed. Wiley- Liss, New Jersey.

Hotton, F., Kleiner, S., Bollaert, A. Twiesselman, F. Le rocher des Neanderthaliens de Spy. Etude radio-anatomique. *J. Belge Radiol.* 59 (1976):39–50.

Hoyte, D. 1997. Growth of the cranial base. Chapter 11, *In* A. D. Dixon, et al., eds. *Fundamentals of Craniofacial Growth*. CRC-Press, Florida.

Hublin, J.-J., Spoor, F., Braun, M., Zonneveld, F., and Condemi, S. 1996. A late Neanderthal associated with Upper Palaeolithic artefacts. *Nature* 381(6579):224–226.

Jeffery, N., and Spoor, F. 2002. Brain Size and the Human Cranial Base: A Prenatal Perspective. *American Journal of Physical Anthropology* 118(4):324–340.

Jian, F., and L. Hongnian. 2006. Beam-hardening correction method based on original sinogram for X-CT. Nuclear Instruments and Methods in Physics Research Section A: Accelerators, Spectrometers, Detectors and Associated Equipment 556:379–385.

Jiménez-Castellanos, J. 1981. *Fundamentos morfológicos de la tomografía axial computarizada cráneo-encefálica*. Secretariado de Publicaciones de la Universidad de Sevilla.

Jungers, W.L. & Minns, R.J. 1979. Computed Tomography and biomechanical analysis of fossil long bones. *Am J Phys Anthrop*, 50 p 285–290.

Kalvin, A., Dean, D., and Hublin, J.-J. 1995. Reconstruction of human fossils. *Computer Graphics and Applications, IEEE* 15:12–15.

Kappers, C. 1936. The endocranial casts of the Ehringsdorf and *Homo soloensis* skulls. *Journal of Anatomy*. 71(1):61–76.

Keith, A. 1931. *New discoveries relating to the antiquity of man*. London: W. W. Norton.

Ketcham, R., and Carlson, W. 2001. Acquisition, optimization and interpretation of X-ray computed tomographic imagery: applications to the geosciences. *Computers & Geosciences* 27:381–400.

Kijewski, P. et al., Correction for beam hardening in computed tomography, Med. Phys., 5(3), pp. 209–214, May/Jun. 1978.

Kuroe, K., Rosas, A., and Molleson, T. 2004. Variation in the cranial base orientation and facial skeleton in dry skulls sampled from three major populations. *The European Journal of Orthodontics* 26:201–207.

Laitman, J. 2004. New Eyes for Old Bones: The Power of Virtual Study of the Fossil Hominid From Eliye Springs, Kenya, by CT-Based 3D Reconstruction. *The Anatomical Record (Part B: New Anat.)* 278B:2–3.

Le Floch-Prigent, P. 1989. Scannographie du crâne de Pétralona: coupes systématiques dans les trois plans. I: Résultats morphologiques. *Comptes rendus de l'Académie des sciences.* Série 2, 309:1855–1862.

Legoux, P., 1966. Détermination de l'age dentaire de fossiles de la lignée humaine. Libraire Maloine S.A., Paris.

Lieberman, D. 1995. Testing hypotheses about recent human evolution from skulls. *Current Anthropology.* **36,** 159–197.

Macho, G., and Thackeray, J. 1992. Computed tomography and enamel thickness of maxillary molars of Plio-Pleistocene hominids from Sterkfontein, Swartkrans, and Kromdraai (South Africa): An exploratory study. American *Journal of Physical Anthropology* 89:133–143.

MacLeod, C., Zilles, K., Schleicher, A., Rilling, J., and Gibson, K. 2003. Expansion of the neocerebellum in Hominoidea. *Journal of Human Evolution* 44(4):401–429.

Mafart, B., Guipert, G., Lumley, M.A., and Subsol, G. 2004. Three dimensional computer imaging of hominid fossils: a new step in human evolution studies. *Canadian Association of Radiologists Journal* 55(4):264–270.

Manzi, G., Bruner, E., Caprasecca, S., Gualdi, G., and Passarello, P. 2001. CT-scanning and virtual reproduction of the Saccopastore Neanderthal crania. *Rivista di Antropologia* 79:61–72.

Martinez, I., Rosa, M., Arsuaga, J. -L., Jarabo, P., Quam, R., Lorenzo, C., Gracia, A., Carretero, J. -M., Bermudez de Castro, J. -M., and Carbonell, E. 2004. Auditory capacities in Middle Pleistocene humans from the Sierra de Atapuerca in Spain. *Proceedings of the National Academy of Sciences* 101(27):9976–9981.

Mazurier, A., Volpato, V., and Macchiarelli, R. 2006. Improved noninvasive microstructural analysis of fossil tissues by means of SR-microtomography. *Applied Physics A: Materials Science & Processing* 83:229–233.

Minugh-Purvis N. 1988. *Patterns of craniofacial growth and development in Upper Pleistocene hominids.* PhD Dissertation (anthropology) University of Pennsylvania.

Minugh-Purvis, N., Radovc, J., and Smith, F. 2000. Krapina 1: A Juvenile Neanderthal From The Early Late Pleistocene Of Croatia. *American Journal Of Physical Anthropology* 111:393–424.

Neeser, R. 2007. *A Comparison of Statistical and Geometric Reconstruction Techniques: Guidelines for Correcting Fossil Hominin Crania.* PhD Thesis. Department Of Computer Science, Faculty Of Science, University Of Cape Town: pp. 192.

Neubauer, S., Gunz, P., Mitteroecker, P., and Weber, G. 2004. Three-dimensional digital imaging of the partial *Australopithecus africanus* endocranium MLD 37/38. *Canadian Association of Radiologists* 55:271–278.

Olejniczak, A. and Grine, F. 2005. High-resolution measurement of Neandertal tooth enamel thickness by micro-focal computed tomography. S Afr J Science 101 p 219–220

Olejniczak, A. and Grine, F. 2006. Assessment Of The Accuracy Of Dental Enamel Thickness Measurements Using Microfocal X-Ray Computed Tomography. *The Anatomical Record Part A*, 288a:263–275pp.

Olejniczak, A.J., Grine, F.E., and Martin, L.B.. 2007a. Micro-computed tomography of the post-canine dentition: methodological aspects of three-dimensional data collection. In: (Bailey SE, Hublin J-J, Eds.) Dental Perspectives on Human Evolution: State of the Art Research in Dental Anthropology. Springer, Dordrecht, p 103–115.

Olejniczak, A.J., Tafforeau, P., Smith, T.M., Temming, H., Hublin, J.-J. 2007b.Technical note: Compatibility of microtomographic imaging systems for dental measurements. *Am J Phys Anthropol,* 134 p 130–134.

Olejniczak, A., Tafforeau, P., Feeney, R., and Martin, L. 2008. Three-dimensional primate molar enamel thickness. *Journal of Human Evolution* 54:187–195.

Ponce De Leon, M., and Zollikofer, C. 1999. New Evidence from Le Moustier 1: Computer-Assisted Reconstruction and Morphometry of the Skull. *The Anatomical Record* 254:474–489.

Ponce de León, M. 2002. Computerized paleoanthropology and Neanderthals: The case of Le Moustier 1. *Evolutionary Anthropology* suppl 1:68–72.

Poza-Rey, E.M., and Arsuaga, J.L. 2009. Reconstitution 3D par Computerized-tomography (CT) et endocrâne virtuel du crâne 5 du site de la Sima de Los Huesos (Atapuerca). "Les Premiers habitants de l'Europe " (Colloque C13, Lisbonne 2006). *L'Anthropologie* 113:211–221.

Price, J.L., Molleson, T.I. 1974. A radiographic examination of the left temporal bone of Kabwe man, Broken Hill mine, Zambia. *J Archaeol Sc* 1, p 285–289.

Prossinger, H., Seidler, H., Wicke, L., Weaver, D., Recheis, W., Stringer, C., and Mullër, G. 2003. Electronic removal of encrustations inside the Steinheim cranium reveals paranasal sinus features and deformations, and provides a revised endocranial volume estimation. *The Anatomical Record* (Part B: New Anat) 273b:132–142.

Radon, J. 1917. Ueber die Bestimmung von Funktionen durch ihre Integralwerte längs gewisser Mannigfaltigkeiten. *Ber. Verh. Sächs. Akad.* 69 p 262–277.

Rae, T., and Koppe, T. 2000. Isometric scaling of maxillary sinus volume in hominoids. *Journal Of Human Evolution* 38:411–423.

Rae, T., and Koppe, T. 2002. 3D imaging and measurement in studies of cranial pneumatisation. In: Mafart B, Delingette H, editors. *Three-dimensional imaging in palaeoanthropology and prehistoric archaeology*. Oxford: British Archaeological Report International Series 1049. p. 11–6.

Raptopoulos, V., D. Schellinger, and S. Katz. 1978. Computed Tomography of Solitary Pulmonary Nodules: Experience with Scanning Times Longer than Breath-Holding. Journal of Computer Assisted Tomography 2:55–60.

Recheis, W., Weber, G., Schafer, K., Knapp, R., Seidler, H., and zur Nedden, D. 1999. Virtual reality and anthropology. *European Journal of Radiology* 31:88–96.

Saban, R. 1984. Anatomie et évolution des veines méningées chez les hommes fossiles. *Memoires de la section sciences n° 11*, E.N.S.B.- C.T.H.S. Paris.

Schoetensack, O. 1908. *Der Unterkiefer des Homo heidelbergensis aus den Sanden von Mauer, bei Heidelberg*, Engelmann, Leipzig.

Schulz, T., and Ingensand, H. 2004. Terrestrial laser scanning- investigations and applications for high precision scanning. In: *Proceedings ofthe "FIG Working Week – The Olympic Spirit in Surveying"*. Athens.

Schwartz, G. 2000. Taxonomic and functional aspects of the patterning of enamel thickness distribution in extant large-bodied hominoids. *American Journal of Physical Anthropology* 111:221–244.

Seidler, H., Falk, D., Stringer, C., Wilfing, H., Muller, G., zur Nedden, D., Weber, G., Reicheis, W., and Arsuaga, J.L., 1997. A comparative study of stereolithographically modelled skulls of Petralona and Broken Hill: implications for future studies of middle Pleistocene hominid evolution. *Journal of Human Evolution* 33:691–703.

Skinner, M., and Sperber, G. 1982. *Atlas of radiographs of early man*. New York: Alan R. Liss. p 1–346.

Smith, F., and Ranyard, G. 1980. Evolution of the supraorbital region in Upper Pleistocene hominids from South-Central Europe. *American Journal of Physical Anthropology*. 53, 589–610.

Spoor, F., and Zonneveld, F. 1995. Morphometry of the primate bony labyrinth: a new method based on high-resolution computed tomography. *Journal of Anatomy* 186:271–286.

Spoor, F., and Zonneveld, F. 1998. Comparative review of the human bony labyrinth. *American Journal of Physical Anthropology* 107:211–251.

Spoor, F., and Zonneveld, F. 1999. Computed Tomography-based Three-Dimensional imaging of Hominid Fossils: Features of the Broken Hill 1, Wadjak 1, and SK 47 Crania., p. 207–226, *In* Koppe, T. et al., eds. *The paranasal sinuses of higher primates: development, function, and evolution*. Quintessence Publishing Co, Chicago.

Spoor, F., Zonneveld, F., and Macho, G. 1993. Linear measurements of Cortical Bone and Dental Enamel by Computed Tomography: Applications and Problems. *American Journal of Physical Anthropology* 91:469–484.

Spoor, F., Wood, B., and Zonneveld, F. 1994. Implications of early hominid labyrinthine morphology for evolution of human bipedal locomotion. *Nature* 369:645–648.

Spoor, F., Jeffery, N., and Zonneveld, F. 2000a. Using diagnostic radiology in human evolutionary studies. *Journal of Anatomy*. 197:61–76.

Spoor, F., Jeffery, N., and Zonneveld, F. 2000b. Imaging skeletal growth and evolution. Chapter 6. *Development, growth, and evolution*. P. O'Higgins and M. J. Cohn, Elsevier Science & Technology Books. 6:288.

Spoor, F., Hublin, J.J., Braun, M., and Zonneveld, F. 2003. The bony labyrinth of Neanderthals. *Journal of Human Evolution* 44:141–165.

Suwa, G., Asfaw, B., Kono, R., Kubo, D., Lovejoy, C., White, T. 2009a. The *Ardipithecus ramidus* Skull and Its Implications for Hominid Origins. *Science* 326, 68

Suwa, G., Kono, R., Simpson, S., Asfaw, B., Lovejoy, C., White, T. 2009b Paleobiological Implications of the *Ardipithecus ramidus* Dentition. *Science* 326, 69

Symington, J. 1916. Endocranial Casts And Brain Form: A Criticism Of Some Recent Speculations. *Journal of Anatomy and Physiology*. 50(2):111–130.

Tate, J.R. & Cann, C.E. 1982 High resolution Computed Tomography for the comparative study of fossil and extant bone. *Am J Phys Anthrop*, 58 p 67–73.

Thompson, J. L. and Illerhaus, B. 1998. A new reconstruction of the Le Moustier 1 skull and investigation of internal structures using 3-D-micro-CT data. *Journal of Human Evolution* 35 (6):647–665.

Tobias, P. 2001. Re-Creating Ancient Hominid Virtual Endocasts By Ct-Scanning. *Clinical Anatomy* 14:134–141.

Urquiza, R., Botella, M., and Ciges, M. 2005. Study of a temporal bone of *Homo heildelbergensis*. *Acta Oto-Laryngologica* 125(5):457–463.

Van Tiggelen, R. 2001. Since 1895, Orthopaedic Surgery Needs X-Ray Imaging: A Historical Overview From Discovery To Computed Tomography. *Journal Belge de Radiologie - Belgisch Tijdschrift voor Radiologi* 84:204–213.

Van Tiggelen, R. 2002. In search for the third dimension: from radiostereoscopy to three-dimensional imaging. *Journal Belge de Radiologie- Belgisch Tijdschrift voor Radiologi* 85:266–70.

Van Tiggelen, R., and Pouders, E. 2003. Ultrasound and computed tomography: Spin-offs of the world wars. *Journal Belge de Radiologie- Belgisch Tijdschrift voor Radiologi* 86:235–241.

Walker, A. & Leakey, R. 1993. The skull. In A. Walker & R.Leakey, Eds. *The Nariokotome Homo erectus Skeleton*, pp. 64–94. Harvard: Harvard University Press.

Webb, S. 1990. *From the Watching of Shadows: the Origins of Radiological Tomography*. Bristol; New York. Adam Hilger, 347 pp.

Webb, S. 1992. Historical experiments predating commercially available computed tomography. *Br J Radiol* 65:835–837.

Weber, G. 2001. Virtual anthropology (VA): A call for Glasnost in Paleoanthropology. *The Anatomical Record (New Anat)* 265:193–201.

Weber, G., Kim, J., Neumaier, A., Magori, C., Saanane, C., Recheis, W., and Seidler, H. 2000. Thickness Mapping of the Occipital Bone on CT-data: a New Approach Applied on OH 9. *Acta Anthropologica Sinica* 19 (Supplement):37–46.

Weber, G., Schäfer, K., Prossinger, H., Gunz, P., Mitteröcker, P., and Seidler, H. 2001. Virtual Anthropology: The Digital Evolution in Anthropological Sciences. *Journal of Physiological Anthropology and Applied Human Science* 20:69–80.

Weidenreich, F. 1943. The skull of *Sinanthropus pekinensis*, a comparative study on a primitive hominid skull. *Pal Sin new series D 10*, p 1–291.

Weidenreich, F. 1946. *Apes, giants, and man*. Chicago: The University of Chicago Press (Science Education)

Weiglein AH. 1999. Development of the paranasal sinuses in humans. In: Koppe T, Nagai H, Alt KW, editors. *The paranasal sinuses of higher primates*. Munich: Quintessence. p 35–50.

Weinmann, J.P., and Sicher, H. 1955. *Bone and bones: Fundamentals of bone biology,* 2nd ed. London: Kimpton

Wilhite, R. 2003. Digitizing Large Fossil Skeletal Elements for Three-Dimensional Applications. *Palaeontologia Electronica* 5(1):10pp. http://palaeo-electronica.org/paleo/2002_2/scan/issue2_02.htm

Wind, J. 1984. Computerised x-ray tomography of fossil hominid skulls. *American Journal of Physical Anthropology* 63:265–282.

Wind, J., and Zonneveld, F. 1985. Radiology of fossil hominid skulls. In Alan R. Liss, Inc. *Hominid Evolution: Past, Present and Future*. 427–436pp.

Wind, J., and Zonneveld, F. 1989. Computed Tomography of an *Australopithecus* Skull (Mrs Ples): A new Technique. *Naturwissenschaften* 76:325–327.

Wu, X., Schepartz, L., Falk, D., and Liu, W. 2006. Endocranial cast of Hexian *Homo erectus* from South China. *American Journal of Physical Anthropology* 130:445–454.

Wu, X., Schepartz, L., and Norton, C. 2010. Morphological and morphometric analysis of variation in the Zhoukoudian *Homo erectus* brain endocasts. *Quaternary International*. Volume 211, Issues 1–2, 4–13 pp

Zollikofer, C. 2002. A computational approach to Paleoanthopology. *Evolutionary Anthropology* suppl 1:64–67.

Zollikofer, C., Ponce de León, M., Martin, R and Stucki, P. 1995. Neanderthal computer skulls. *Nature* 375:283–285.

Zollikofer, C., Ponce de León, M., and Martin, R. 1998. Computer-assisted paleoanthropology. *Evolutionary Anthropology: Issues, News, and Reviews* 6(2):41–54.

Zollikofer, C., Ponce de León, M, Lieberman, D., Guy, F., Pilbeam, D., Likius, A., Mackaye, H., Vignaud, P., and Brunet, M. 2005. Virtual cranial reconstruction of *Sahelanthropus tchadensis*. *Nature* 434(7034):755–759.

Zonneveld, F. 1994. A decade of clinical three-dimensional imaging: a review. Part III: Image analysis and interaction, display options, and physical models. *Investigative Radiology* 1994;29 (7):716–725.

Zonneveld, F., and Fukuta, K. 1994. A decade of clinical three-dimensional imaging: a review. Part II: clinical applications. *Investigative Radiology*; 29 (5):574–589.

Zonneveld, F., and Wind, J. 1985. High-resolution Computed Tomography of fossil hominid skulls: a new method and some results. *Hominid Evolution: Past, Present and Future*: 427–436.

Zonneveld, F., Spoor, F., and Wind, J. 1989. The use of the CT in the study of the internal morphology of hominid fossils. *Medicamundi* 34:117–128.

Chapter 8
From a Skeleton to a 3D Dinosaur

Stefan Stoinski

Introduction

In 2004, researchers from the Institute of Computer Vision and Remote Sensing of the Technical University of Berlin joined the Research Group FOR 533 "Biology of the Sauropod Dinosaurs: the Evolution of Gigantism" of the German Research Foundation. The interdisciplinary and international research group's goal was to explain how sauropods could grow to their exceptional body size, which eclipses all other terrestrial vertebrate groups by at least an order of magnitude. Most of theories advanced during the course of the FOR 533's research involved body volumes and masses. However, in order to have realistic volume and mass estimates, new modeling methods were needed. This chapter gives a short overview of previous methods for mass estimation of extinct animals, and describes both photogrammetry of dinosaur skeletons and a method for estimating body mass of sauropod dinosaurs using body volumes derived from laser-scanning (Gunga et al. 1999, 2007, 2008; Bellmann et al. 2005).

Why Estimate Body Mass?

Body size is one of the most fundamental attributes of any organism, being linked to the genetically determined bauplan, life history, and ecology (Clutton-Brock et al. 1980; Peters 1983; Schmidt-Nielsen 1984; Alexander 1998; Hunt and Roy 2005; Makarieva et al. 2005; Bonner 2006; Bates et al. 2009). A lot of physiological data can be derived from an estimate of body mass, using allometric formulae (Schmidt-Nielsen 1984, 1997; Calder 1996). For the largest terrestrial animals of all times – the sauropod dinosaurs – knowing this data means holding key information for

S. Stoinski
Computer Vision a Remote Sensing, Technical University Berlin, FR 3 -1; Franklinstr. 28/29, 10587 Berlin, Germany
e-mail: stoinski@fpk.tu-berlin.de; http://www.cs.tu-berlin.de

A.M.T. Elewa (ed.), *Computational Paleontology*,
DOI 10.1007/978-3-642-16271-8_8,

understanding their biology and gigantism. However, very different body mass estimations can be found in literature. For example, published mass estimates for *Giraffatitan brancai* (Janensch, 1914) (until recently considered to belong to *Brachiosaurus* Riggs, 1903 as *B. brancai*, see Taylor 2009) vary between slightly over 23 tons (Taylor 2009) and 85 tons (Colbert 1962). For *Plateosaurus*, values between 279 kg (Henderson 2006) and 6 tons (Sander 1992) have been published. Based on such mass estimates, for example, inferences on locomotion capabilities, energy requirements, energy balance, and ultimately metabolic rate can be drawn (see Sander et al. 2010). Many important parameters can be calculated via allometric formulae (Schmidt-Nielsen 1984, 1997; Calder 1996), increasing the need for accurate estimates.

Previous Mass Estimation Methods

There are various methods to estimate body mass of an extinct animal. A review is given in Sander et al. (2010). Methods can be separated into two groups: some uses biomechanical or scaling approaches to estimate the body volume. For example, it is possible to extrapolate body mass based on using long bone circumferences (Anderson et al. 1985, formula corrected by Alexander 1989). Usually more accurate are methods that reconstruct the body volume and, from it, calculate body mass. A classic method, going back to Archimedes, is weighing a dinosaur scale model in air and water (e.g., Gregory 1905; Alexander 1989; Mazzetta et al. 2004). But minimal mistakes in the model can significantly influence the calculated body mass. Digital 3D reconstructions, or purely mathematical models (e.g., thorough 3D mathematical slicing; Henderson 1999), are sometimes based on reconstruction drawings (e.g., Henderson 1999, 2006; Seebacher 2001; Taylor 2009). For these drawings, an unofficial standard was developed by Robert Bakker (e.g., Bakker 1986) and especially Gregory Paul (e.g., Paul 1987, 1997, 2000), among others. Usually, skeletons are drawn in lateral view, in a rapid walking or running pose, with a black outline showing a suggested body outline. Some researchers provide drawings in which they indicate what parts of the animal are based on fossil specimens, and what parts are reconstructed (e.g., Scott Hartmann, www.skeletaldrawing.com). These drawings are often called "rigorous" drawings. However, these drawings can be significantly inaccurate (Mallison 2010b), and 3D models based on them (e.g., Henderson 1999, 2006; Seebacher 2001) thus arrive at inaccurate weights.

During the 1990s the department of Photogrammetry and Cartography of the Technical University Berlin (now Department of Computer Vision and Remote Sensing) began measuring dinosaurs, and developed a novel method to acquire data for body mass estimates. The skeleton of *Giraffatitan brancai* mounted in the Museum für Naturkunde Berlin was measured photogrammetrically (Gunga et al. 1995). The analysis of that measurement resulted in a non-scaled grid model (Fig. 8.1). Paper copies of several parts and of the whole animal were produced and a physiologist drew outlines – the special contours of the body surface – around the cross sectional area. After that, the skeleton was portioned and geometric

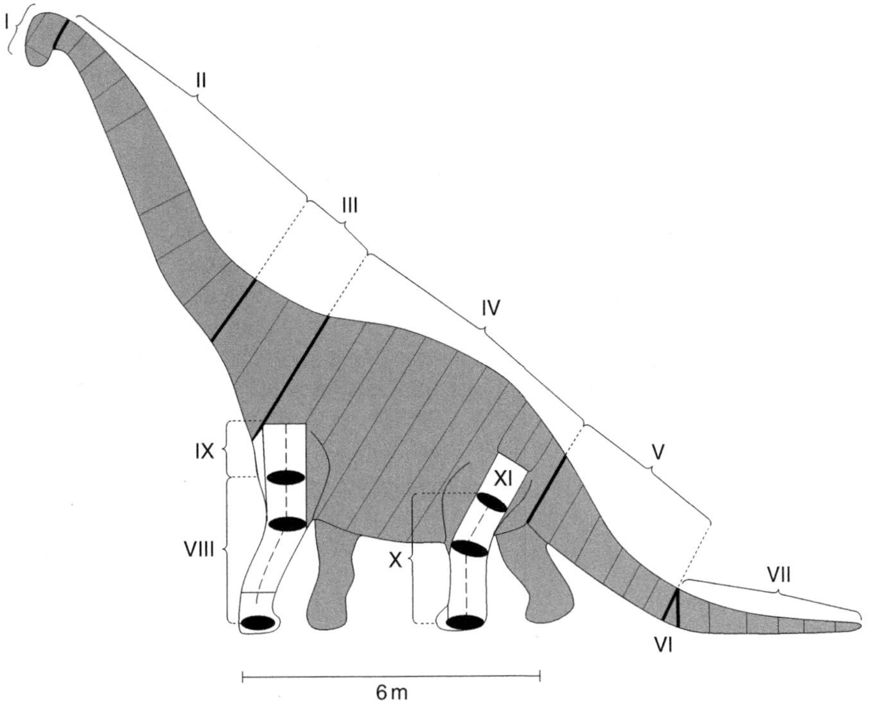

Fig. 8.1 Drawing of *Giraffatitan* model based on photogrammetric mesurements. On the basis of this drawing's sectioning geometric primitives were used to calculate the total body volume

primitives (cylinders, cones and spheres) were fit into the contours (Fig. 8.1). Then, the volume of every part was calculated, and the resulting values were summarized.

Here, I describe both the photogrammetry method, and laser scan based reconstructions (e.g., Gunga et al. 1999, 2007, 2008; Bellmann et al. 2005), the latter similar to the method independently developed by Bates et al. (2009).

Methods

Data Collection

Photogrammetry

Well preserved fossil skeletons are the ideal basis for 3D reconstructions. Many of these are mounted in museum exhibitions, resulting in limited accessibility. To reconstruct the animal's in vivo shape, it is necessary to create 3D-point clouds of the skeletons. A method that allows measuring skeletons in one piece and at one go, without dismounting, is the use of a photogrammetric camera (Gunga et al. 1995; Wiedemann et al. 1999). For this method, a large number of photographs of the skeleton are captured with a fixed focal distance from different positions. Each photo is overlaid with a grid, based on a thin glass sheet over the film footage, called Réseau-grid. This allows removing the edge distortion of each individual photograph. The viewing direction (the direction the lens points in) should ideally be perpendicular to the basis – the line between the two points the images are taken from – which should also be parallel to the object's front. Using an analytical stereo photogrammetric instrument, e.g., the Kern DSR 11, it is possible to reconstruct 3D-points out of metric 2D-images arithmetically (see, e.g., Wiedemann et al. 1999). The main advantage of this method is that many images can be captured in a very short time. However, the analysis of the stereo image pairs is very time consuming, and the use of only certain distinctive points seen in both images, instead of a full "skin paint" is a further disadvantage. Additionally, if insufficient images are taken, and parts of the mount thus not sufficiently sampled, this will often be recognized only during the 3D reconstruction, requiring another visit to the mount. The method produces, as a final result, a grid model of the measured skeleton.

Laser Scanning

Laser scanning of mounted skeletons has many advantages over photogrammetry, being the prime method for measuring in detail objects consisting of many individual parts, e.g., mounted skeletons. The advantages of laser scanning are the enormous number of points of the surface that can be captured during the scan process, the small amount of time consumed due to the short capture time, and the ability to see the growing point cloud during the measurement, which gives an immediate feedback whether the scan is set up correctly and will deliver the desired result. The last point is helpful for recognizing holes, for example, caused by occlusions, which can immediately be filled by additional scans from other viewpoints. This makes repeat visits to the museum exhibitions unnecessary, in the case of international or even intercontinental travel an advantage that cannot be overestimated.

For our work as part of the German Science Foundation Research Unit 533 "Sauropod Biology" we used a S25 laser scanner manufactured by Mensi S.A. This triangulation-type laser-scanner has a measurement range between 0.8 and 25 m and a data acquisition rate of up to 100 points per second. The measurement accuracy orthogonal to range is 0.8–3.8 mm vertically and 0.2–3.4 mm horizontally. The distance accuracy is about 0.2 mm at 4 m distance, and 1.4 mm at 10 m (Boehler et al. 2003). The field of view is about $320^\circ \times 46^\circ$, so that tall objects can be measured quickly.

The principle of the triangulation laser scanner is shown in Fig. 8.2. A laser creates a spotlight on the surface to be scanned, and the reflected light is collected by an acceptor optic and received on a CCD sensor. From the position of the spotlight on the CCD sensor, the measured distance and the fixed basis between

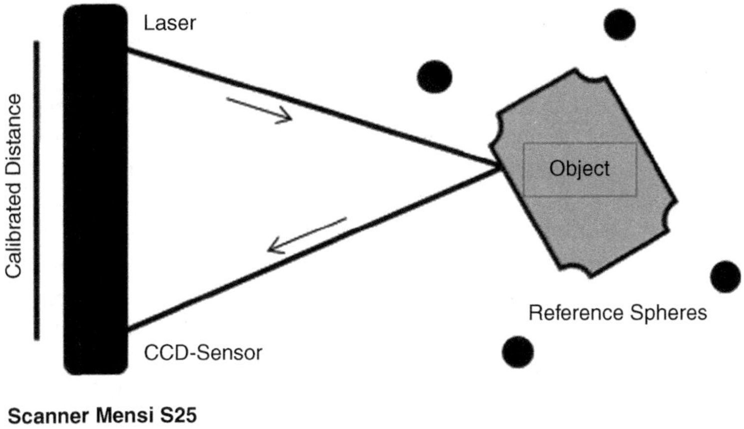

Fig. 8.2 Schematic depiction of the principle of laser scanning

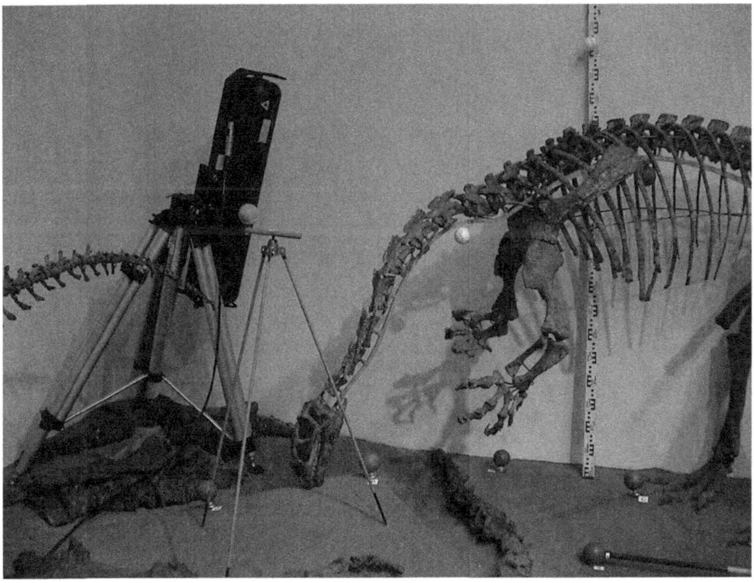

Fig. 8.3 Laser scanning *Plateosaurus engelhardti* GPIT/RE/7288. On the left, the scanner is positioned almost vertical, to capture the side of the mount facing the wall. Note reference objects (styrofoam spheres) of different sizes

the CCD sensor and the laser root the 3D position of the spotlight on the surface is computed.

To scan an object, it is first surrounded by reference objects. These reference objects usually are spheres, because the centre of a sphere can be localized from every point of view (Fig. 8.3). The spheres are numbered, of different size, and their position must not be altered for the duration of all measurements. The reference objects are necessary, because the scanner defines a local coordinate system for each separate scan and viewpoint. All local coordinate systems of all viewpoints have to be transformed into one global coordinate system at the end of the measurement process to create one continuous point cloud of the object. A minimum of three spheres visible in every viewpoint is necessary to perform this affine transformation (defined about three transformations, three rotations, and three scaling parameters).

From every viewpoint, the reference spheres must be measured before the objects, or any part of it. The reference objects have known sizes; we used Styrofoam spheres of 70, 80 and 76.2 mm diameter. Thus it is possible to check the difference between the measured and the real diameter during and after measurement. This gives an indication of the accuracy for the current measurement, and thus of the whole measurement. The measuring grid can be adjusted manually, and starts at a minimum grid density of 0.1 mm. Because simple geometric objects, especially planar objects, can be described with few points, the measuring grid resolution can be reduced, resulting in smaller final file sizes and faster scanning. The growing point cloud can be inspected on screen during the scanning process. This allows immediately spotting holes in the scan, caused by parts of the object being obscured or by faulty scan set-up, so that additional scans from other viewpoints can be taken to fill the gaps.

When all measurements are completed, the registration – the transformation of all local coordinate systems into one global coordinate system– can be performed. The registered point cloud includes all points measured on the object of interest. However, it also includes additional, undesired points, e.g., the ground or visible support elements. These points now have to be deleted (Fig. 8.4).

Because of the irregular surface of the objects, the editing process must be done manually. Compared to the scanning process or the 3D modeling described

Fig. 8.4 Final, registered point cloud of laser scan of *Giraffatitan brancai* in the MFN. On the *left*, all collected data is shown, including undesirable points from the museum building (*black*, *top* and *long lines*), other skeletons and support structures (*black*, *below*) and reference objects. On the *right* is shown the pared down data, depicting only the skeleton without junk data

below, it is the most time consuming part of the whole reconstruction process. The resulting point cloud without junk points (Fig. 8.4) is the final result of the measuring and editing processes. The editing is usually done by using an application in the scanner software, but can also be performed in any other CAD software able to handle point clouds.

Digital 3D Modelling

Once a 3D scan of a mounted skeleton has been produced, the shape of the animal's soft tissues needs to be reconstructed. Our first attempt at this first step for obtaining a reliable body volume estimate, the modeling of 3D body volumes, involved creating the body surface using rotational solids (Bellmann et al. 2005). A juvenile Indian elephant was modeled using the computer aided design (CAD) software AutoCAD. The distinct body sections (limbs, trunk, body, and head) were created separately. Images of elephants were used as cues to get optimal results. The skeleton scan and final 3D-model are shown in Fig. 8.5.

Because of the unrealistic appearance of the elephant model we started searching for new methods and software to create more plausible models. The next models were created using NURBS-modeling. NURBS stands for *non-uniform rational B-splines*, essentially editable 3D curves that allowed us to design shapes freely. The resulting freeform shapes approximated the body surface better than rotational solids. The working steps to produce a body surface from a laser scan were nearly the same as those used in the 1990s on photogrammetric data. First, the outlines of the body shape were drawn in the CAD program around specific positions of the digital skeleton (Fig. 8.6). These contours were discussed with other researchers from relevant fields in the research group, and if necessary improved. The latest

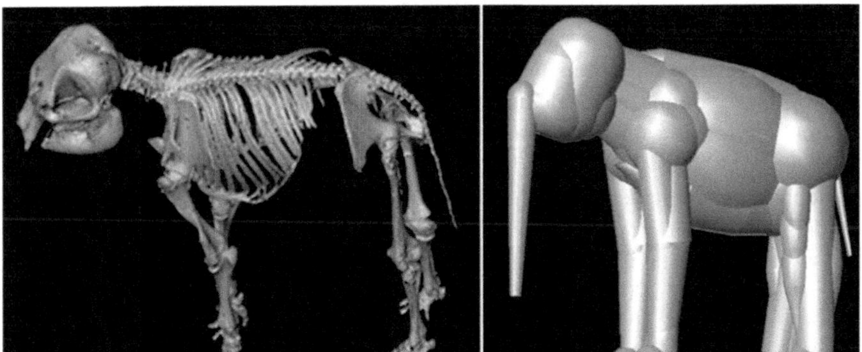

Fig. 8.5 Skeleton of juvenile *Elephas maximus*, and CAD model from geometric primitives (see Bellmann et al. 2005)

Fig. 8.6 *Giraffatitan brancai*, laser scan data and CAD model. On the left, contour curves have been drawn around the point cloud, on the right those for the body have been used to create a 3D surface

research results were taken into account, for example, on neck shape or shoulder girdle arrangement and position (Wedel 2005; Schwarz-Wings et al. 2007; Schwarz-Wings 2009). In the next step, NURBS freeform shapes were calculated from the contours (Fig. 8.7b), and the resulting 3D-model was reviewed by the experts again.

The advantage of NURBS-modeling is the ability to realize changes and corrections quickly. Contour curves are easy to edit, so that small and even large alterations to the 3D shapes can be effected in minutes. Once the model had been finalized, its volume was calculated, and the animal's body mass was estimated by multiplying it with a specific density value. The specific density is the ratio between body mass and volume (here given in kilogram per liter). In most terrestrial vertebrates, specific density is close to 1 kg/L. This value was used in our earlier studies (Gunga et al. 1995, 1999, 2002). Later, research showed that sauropod dinosaurs had a heavily pneumatized skeleton and large pulmonary air sacs (e.g., Perry 2001; Wedel 2003a, b, 2005; Schwarz and Fritsch 2006), so that a much lower specific density of only 0.8 to 0.9 kg/L seems to be more likely. In our recent studies (Gunga et al. 2007, 2008), a value of 0.8 kg/L was used, as suggested as the highest likely value by Wedel (2005).

Next, the sizes of interior organs (heart, liver, etc.) and organ systems (e.g., digestive tract) were estimated using allometric formulae. To check the accuracy of

Fig. 8.7 CAD models of *Plateosaurus engelhardti* from Gunga et al. 2008

the initial model, 3D shapes of the organs and organ systems were placed into the body cavity of the model. If the fit was poor (organs too large or too small), the model had to be adapted. While this process is highly speculative, because especially air sac size is hard to estimate, gross errors can be avoided. The models of the basal sauropodomorph *Plateosaurus engelhardti* von Meyer, 1837 (Gunga et al. 2007) are a good example. Initially, a large model was created (Fig. 8.7). The reconstruction of the internal organs showed that this model was much too "fat". Therefore, a slimmer version of *Plateosaurus* was created, resulting in a good fit of the calculated internal organs (Gunga et al. 2008; Fig. 8.7).

Tests of Method Accuracy

In order to test the accuracy of our volume reconstruction and estimation methods, two extant animals were modeled, for which the body mass at death was known. A mounted (padded) rhinoceros and the skeleton of a juvenile Indian elephant were scanned in 2004.

The rhinoceros is on exhibit in the Senckenberg Naturhistorische Sammlungen Dresden, Germany. We needed two days to scan this rhino, which is roughly 3 m long, 0.8 m wide and 1.5 m high. Fifteen reference spheres and 16 scanner positions were used. The resulting point cloud includes more than 570,000 points, and the accuracy after registration was better than 3 mm (Fig. 8.8). According to museum records, the life weight of the rhinoceros was 1,050 kg. As the computed volume of the NURBS surface based on the scan is 0.914 m^3, a specific density of 1.15 kg/l for this specimens was calculated (Bellmann et al. 2005).

The skeleton of the juvenile Indian elephant (*Elephas maximus* Linnaeus, 1758) mentioned above is mounted in the Zoological Museum of Copenhagen, Denmark (Fig. 8.5). The skeleton was scanned within two days by using seven scanner positions and 15 reference points. The registration accuracy of the more than 920.000 scanned points is better than 1 mm. Afterwards, the body surface area

Fig. 8.8 Stuffed (padded) rhinoceros with reference objects prepared for scanning (*top*), and meshed, unedited point cloud (*bottom*)

was modeled in AutoCAD (Fig. 8.5). The body volume of the 1.7 m long, 0.7 m high and 1.5 m wide was calculated at 0.622 m^3. The Indian elephant had an estimated body mass of 715 kg, using the specific density of 1.15 obtained from the padded rhino specimen. This mass compares well to the known life mass of the elephant of 850 kg. Our method underestimates its life mass by 16%.

Limitations

All methods for estimating body volumes and body masses of dinosaurs have certain sources of error. These lead to differences between a mass estimate based solely on a skeleton and a value derived from measuring a living animal. The two major sources of error are the uncertainties involved in mounting the skeleton and the error introduced by digitally reconstructing the body surface.

When using laser scanning of complete mounts, the first sources of error is the anatomical quality of the skeleton mounts. Also, the authenticity of the skeletons needs to be carefully checked. Most of the mounted skeletons of dinosaurs are not based on the complete skeleton of one individual. It is standard practice to reconstruct some parts of the skeleton in plaster, or incorporate bones from other individuals into the mount. While skeletons mounted in scientific institutions are usually a good approximation of the skeleton of the living animal, they still require careful scrutiny, especially with regards to the possible mobility in the joints and the overall shape of the animal. Bones may be deformed or misaligned, and technical problems of mounting may require unrealistic bone positions-errors that will be faithfully incorporated into the laser scans. If errors are obvious, it is possible to section the point cloud and manually edit the position of the parts, to improve on the mount. More troublesome are large-scale plaster "improvements" that may go unnoticed, and induce massive errors in the final models.

As mentioned, the other large source of error is the uncertainties in the reconstructing of the soft tissues. This is common to all mass estimation methods based on the reconstruction of soft parts (Alexander 1989; Henderson 1999; Seebacher 2001), in opposition to methods based on measurements derived directly from the bones (e.g., Anderson et al. 1985). Because there are no living animals the size of the large sauropods, and only very few taxa (elephants, arguably rhinoceroses) the size of an "average" sauropod, it is hard to estimate how muscular the extinct giants were. For example, Persons (2009) reported that the size of the caudofemoralis muscle, the main retractor of the limb in most dinosaurs (Gatesy 1990) is usually reconstructed at roughly half the size it should be, if extant reptiles are used as a guide (see also Mallison [2011]). Thus, it is usually a good idea to create several models that differ in the amount of soft tissues, as was done for *Plateosaurus* (Gunga et al. 2007).

Results and Discussion

Laser-Scans of Skeletons

In the course of the FOR 533 research project we scanned a number of sauropod dinosaurs, selecting mounts with an emphasis on separation in time and in the phylogenetic tree. Additionally, several other dinosaur and fossil mammal skeletons were scanned, but will not be discussed here.

At the Museum für Naturkunde Berlin several sauropod mounts, *Giraffatitan brancai* (Janensch, 1914), *Dicreaosaurus hansemanni* Janensch, 1914 and a cast of *Diplodocus carnegii* Marsh 1878, were measured via photogrammetry. A laser scanner was used to measure the skeletons of *Dicreaosaurus* and *Diplodocus* for the first time in 1997. The measurement accuracy of the scans was better than 5 mm and more than 800,000 object points for *Diplodocus* and more than 500,000 points for *Dicreaosaurus* were measured from 21 and 8 viewpoints, respectively. In 2007 the dinosaur hall at the Natural History Museum Berlin was reopened after the building was renovated and the exhibit redesigned. The sauropods were re-mounted under supervision of members of our research group. For comparison with the old mount the skeleton of *Giraffatitan* was measured again in 2008, this time using laser scanning. We needed 1 day and seven viewpoints to scan the skeleton with an overall accuracy better than 5 mm. The accuracy of the long neck and the head was less than 10 mm, due to the greater distance from the scanner, which could only be set up on the floor of the exhibition hall. It would have taken a raised platform, bringing the scanner closer to the neck, to achieve the same accuracy as for the limbs and trunk.

Previously, we had performed our first laser scan of a dinosaur skeleton in summer 2004. The basal sauropodomorph *Plateosaurus engelhardti* (GPIT/RE/7288), a nearly complete individual mounted in the Institute for Geoscience of the University of Tübingen, Germany, was measured at an accuracy better than 3 mm. The mount is roughly 4.3 m long, 1.9 m high, and 0.8 m wide. We needed three days for measurement of more than 1.1 million points, using 10 scanner positions and 17 reference spheres.

In November 2004 the skeleton of the basal macronarian sauropod *Atlasaurus imelakei* Monbaron et al., 1999 mounted in the Ministry of Energy and Mining Rabat (Morocco) was measured. It was very difficult to find good scanner positions, because the skeleton nearly filled out the space of the exhibition hall. The scanning took 4 days, producing more than 660,000 points from 16 scanner positions, with 23 reference points.

In 2005 we had a special opportunity to measure a wall-mounted *Camarasaurus* sp. Cope, 1877 skeleton, as well as free-standing skeletons of *Allosaurus fragilis* Marsh, 1877a and *Stegosaurus* sp. Marsh, 1877b at the Sauriermuseum Aathal, Switzerland. Only one scanner position without reference spheres was used to scan the skeletons of both *Camarasaurus* and *Allosaurus*. The skeleton of *Stegosaurus* was scanned from two scanner positions, also without reference points. The registration of the point clouds had to be performed using their own points, attempting to manually fit the two point clouds as well as possible. This method is less accurate than using reference points, and more difficult to perform without special software.

In May 2005 we ran an extensive measurement campaign in China, at the Beijing Natural History Museum and at the Dinosaur Museum of Zigong. In Beijing we scanned five skeletons. Of special interest were the skeletons of the basal sauropodomorph *Lufengosaurus huenei* Young, 1941 and of the basal eusauropod *Mamenchisaurus jingyanensis* Zhang et al., 1998, which is extremely long-necked. The entire hall containing the dinosaur skeletons was scanned in 7 days using 8 viewpoints and 15 reference points. The registration accuracy was better than 4 mm. At the Zigong Dinosaur Museum, we were able to scan a total of 10 skeletons. These include five sauropod skeletons: a juvenile and an adult of the basal sauropod *Shunosaurus lii* (Dong et al. 1983), an adult *Mamenchisaurus hochuanensis* Young and Zhao, 1992 , an adult of *Omeisaurus tianfuensis* (He et al. 1984), which is closely related to *Mamenchisaurus*, and a skeleton of *Datousaurus bashanensis* (Dong and Tang 1984), a sauropod of uncertain systematic affinities. Twenty viewpoints and 42 reference points were needed to create one point cloud encompassing all dinosaurs in the hall. The registration accuracy was better than 14 mm.

Volumetric Models

To reconstruct body volumes we initially used CAD-software and rotational solids (Bellmann et al. 2005) as described above. The resulting model of the Indian elephant (Bellmann et al. 2005; Fig. 8.5) is obviously very rough, and does not look realistic. NURBS modelling was first tested on the *P. engelhardti* skeleton from Tübingen, giving more aesthetically pleasing and realistic results. This mount is of high quality, with almost the entire skeleton well articulated (Mallison 2010b), so that we started with an accurate data base. For *P. engelhardti* a slim and a robust model were created. The physiologists reconstructed organs and organ systems, showing that the slim reconstruction provided a better fit (Gunga et al. 2007).

Because of the good results with NURBS-modeling achieved for *Plateosaurus*, we created a revised model of *G. brancai* based on the MFN skeleton. Although this skeleton was remounted from 2005 to 2007, for our work the old, pre-2005 mount was used, so that the old and new CAD methods can be compared directly. The basis was a grid model of *G. brancai* from the analysis of the photogrammetric measurement (Gunga et al. 1995; Wiedemann et al. 1999). The volume of the new 3D model is 47.9 m^3 (Gunga et al. 2008). Interestingly, this model has a volume almost identical to that of a NURBS model created by Mallison (2010a), who used reconstruction drawings and photographs of the mount as base data. Compared to the old volume model of *Giraffatitan* from Gunga et al. (1995) our new reconstruction reduces the animal's volume by nearly 30 m^3. The old volume reconstruction has been created by using geometric primitives (spheres, cylinders), and resulted in a volume of 74.4 m^3 (Wiedemann et al. 1999). The comparison of the body mass estimates for the old and the new models shows an even larger discrepancy. The old mass estimation of 74.4 tons was based on a specific density of 1 kg/L. The new body mass estimation based on a specific density of 0.8 kg/L delivers only 38 tons. Thus, the combination of better modeling techniques and additional research results on the average density of sauropods reduces the estimated weight by over 21 tons. Our results thus converge with those of other methods and researchers, e.g., Mallison and Pfretzschner (2005), who reported a volume of 47 m^3 (equivalent, at a density of 0.8 kg/L, to 39 tons), and Taylor (2009), who calculated 23.3 t (nearly 30 m^3, density of 0.8 kg/L) for the Berlin *Giraffatitan* mount. Interestingly, Janensch already estimated the weight at 40 tons (corresponding to a volume of 50 m^3 if a density of 0.8 kg/L is assumed), based simply on his gut feeling and comparison to elephants (Janensch 1938a, b).

As an added benefit, the digital 3D models can be used for other research tasks as well. For example, the animal's center of mass position can be determined with ease, so that posture and buoyancy can be determined (e.g., Henderson 1999, 2003, 2006; Bates et al. 2009; Mallison 2010a). Furthermore, the models can be sectioned and used for kinetic/dynamic modeling of motions (Mallison 2010a, 2011).

Conclusions

Our experience shows that anatomically plausible NURBS modeling is suitable technique to reconstruct the body surface of extinct animals. Deformations of the entire model to accommodate new anatomical evidence are easy to perform rapidly. Compared to older techniques, NURBS modeling allows the creation of less simplified and thus potentially more accurate results. As a side benefit, the new models are also more aesthetically pleasing and less abstracted than previous attempts, looking more alive.

Our method for estimating dinosaur body masses and volumes has certain advantages over other methods. A key factor is that we used real skeletons, and not idealized models and drawings, as has been done by others (e.g., Alexander 1989; Henderson 1999; Seebacher 2001). This advantage is shared by the method of Anderson et al. (1985), but instead of relying on one or two measurements that as assumed to scale with the entire animal, we use the entire known 3D shape. This distinction is not moot, because many dinosaurs do not scale with even close relatives (Therrien and Henderson 2007). Another advantage is that errors introduced by using scaled models are avoided. If the customary method of weighing a model in air and water is used (Colbert 1962; Alexander 1989; Bakker 1986) a minuscule error in the scale model, e.g., an accidental length increase of only a few percent can result in a multi-ton error in the mass estimate for a sauropod. Our models, in contrast, limit such errors to mistakes and inaccuracies in the starting data (i.e., the mounted skeleton) and misjudgments in the reconstruction of the soft tissues. Furthermore, out models can be transferred electronically without delay all around the world, allowing critical scrutiny by other researchers easily, while physical scale models are less easily distributed. Errors in digital models can be corrected with a few mouse clicks, and new research results can be incorporated with ease.

Acknowledgements I thank all the members of the research unit, especially Martin Sander (University of Bonn, Germany) and Oliver Rauhut (Bavarian State Collection for Paleontology and Geology, Munich, Germany), for their help in selecting the dinosaurs. Further thanks go to the staff of the institutions where we scanned the skeletons, especially for their warm response, patience, and support. In particular, I want to thank the following persons: Clara Stefen (Senckenberg Naturhistorische Sammlungen Dresden, Germany), Mogens Andersen and Per Christiansen (Zoological Museum of the University of Copenhagen, Denmark), Wolf-Dieter Heinrich and Heinrich Mallison (Museum für Naturkunde – Leibniz Institute for Research on Evolution and Biodiversity at the Humboldt-University Berlin, Berlin, Germany), Najat Aquesbi and Mohammed Rochdy (Ministere de l'Energie et des Mines, Rabat, Morocco) and Hans-Jakob "Kirby" Siber and his team (Sauriermuseum Aathal, Switzerland). I also thank the Beijing Museum of Natural History and the Zigong Dinosaur Museum for the excellent cooperation and the very special opportunity to scan in China. The German Research Foundation (DFG) funded this research as part of Research Group 533, to which this is contribution # 106.

References

Alexander, R. M. 1989. Mechanics of fossil vertebrates. Journal of the Geological Society, 146: 41–52.

Alexander R.M. 1998. All-time giants: the largest animals and their problems. Palaeontology, 41: 1231–1245.

Anderson, J.F., A. Hall-Martin and D.A. Russell. 1985. Long-bone circumference and weight in mammals, birds and dinosaurs. Journal of Zoology, 207: 53–61.

Bakker, R.T. 1986. The Dinosaur Heresis. William Morrow, New York. 448 pp.

Bates, K.T., P.L. Manning, D. Hodgetts & W.I. Sellers. 2009. Estimating Mass Properties of Dinosaurs Using Laser Imaging and 3D Computer Modelling. PLoS ONE 4(2): e4532. doi:10.1371/journal.pone.0004532

Bellmann, A., T. Suthau, S. Stoinski, A. Friedrich, O. Hellwich & H.-C. Gunga. 2005. 3D-Modelling of Dinosaurs. In Grün and Kahmen (eds.). 7th Conference on Optical 3-D Measurement Techniques. Vienna.

Boehler, W., M. Bordas & A. Marbs. 2003. Investigating laser scanner accuracy. International Committee for Architectural Photogrammetry Symposium. Turkey. October 2003

Bonner, J.T. 2006. Why Size Matters: From Bacteria to Blue Whales. Princeton University Press. 176 pp.

Calder, W. A. III. 1996. Size, Function and Life History. Dover, Mineola, NY.

Clutton-Brock, T. H., S.D. Albon & P.H. Harvey. 1980. Antlers, body size and breeding group size in the Cervidae. Nature, 285: 565–566.

Colbert, E. H. 1962. The weights of dinosaurs. American Museum Novitates, 2076: 1–16.

Cope, E.D. (1877). On a gigantic saurian from the Dakota epoch of Colorado. Paleontological Bulletin, 25: 5–10.

Dong, Z., Tang, Z., 1984. Note on a new mid-Jurassic sauropod (*Datousaurus bashanensis*) from Sichuan Basin China. Vertebrata PalAsiatica, 15: 307–312.

Dong Z., Zhou S. & Zhang Y. 1983. The dinosaurian remains from Sichuan Basin, China. Palaeontologica Sinica (Series C), 23: 1–145.

Gatesy, S.M. 1990. Caudofemoral musculature and the evolution of theropod locomotion. Paleobiology, 16(2):170–186.

Gregory, W.K. 1905. The weight of the *Brontosaurus*. Science, New Series, 22: 572.

Gunga, H.-C., K. Kirsch, F. Baartz, L. Röcker, W.-D. Heinrich, W. Lisowski, A. Wiedemann & J. Albertz. 1995. New data on dimensions of *Brachiosaurus brancai* and their physiological implications. Naturwissenschaften, 82: 190–192.

Gunga, H.C., K. Kirsch, J. Rittweger, L. Röcker, A. Clarke, J. Albertz, A. Wiedermann, S. Mokry, T. Suthau, A. Wehr, D.H. Wolf & H.P. Schultze. 1999. Body size and body volume in two sauropods from the Upper Jurassic of Tendaguru (Tanzania). Mitteilungen des Museums für Naturkunde der Humboldt Universität zu Berlin. Geowissenschaftliche Reihe, 2: 91–102.
Gunga, H.C., K. Kirsch, J. Rittweger, A. Clarke, J. Albertz, A. Wiedemann, T. Mokry, A. Wehr, D.H. Wolf & H.P. Schultze. 2002. Dimensions of *Brachiosaurus brancai*, *Dicraeosaurus hansemanni* and *Diplodocus carnegii* and their implications for gravitattional physiology. In J. Moravec, N. Takeda and P.K. Singal (eds.). Adaptation Biology and Medicine, 3: 156–169.
Gunga, H.-C., T. Suthau, A. Bellmann, A. Friedrich, T. Schwanebeck, S. Stoinski, T. Trippel, K. Kirsch & Hellwich O. 2007. Body mass estimations for *Plateosaurus engelhardti* using laser scanning and 3D reconstruction methods. Naturwissenschaften, 94: 623–630.
Gunga, H.-C., T. Suthau, A. Bellmann, S. Stoinski, A. Friedrich, T. Trippel, K. Kirsch & O. Hellwich. 2008. A new body mass estimation of *Brachiosaurus brancai* Janensch, 1914 mounted and exhibited at the Museum of Natural History (Berlin, Germany). Fossil Record, 11: 33–38.
He, X., Li, K., Cai, K. & Gao, Y. 1984. *Omeisaurus tianfuensis*—a new species of Omeisaurus from Dashanpu, Zigong, Sichuan. Journal of Chengdu College of Geology, 1984(suppl. 2): 13–32.
Henderson, D.M. 1999. Estimating the masses and centers of masses of extinct animals by 3-D mathematical slicing. Paleobiology, 25, 88–106.
Henderson, D.M. 2003. Tipsy punters: sauropod dinosaur pneumaticity, buoyancy and aquatic habits. Proceedings of the Royal Society London B (Suppl.), 271: 180–S183. doi:10.1098/rsbl.2003.0136
Henderson, D.M. 2006. Burly gaits: centers of mass, stability and the trackways of sauropod dinosaurs. Journal of Vertebrate Paleontology, 26: 907–921.
Hunt, G. & Roy, K. 2005. Climate change, body size evolution and Cope's Rule in deep-sea ostracodes. Proceedings of the National Academy of Sciences, USA, 103:1347–1352.
Janensch, W. 1914. Übersicht über die Wirbeltierfauna der Tendaguruschichten, nebst einer kurzen Charakterisierung der neu aufgeführten Arten von Sauropoden. Archiv für Biontologie, 3(1): 81–110.
Janensch, W. 1938a. Vom Urweltriesen *Brachiosaurus*. Aus der Natur (Der Naturforscher), 15: 114–119.
Janensch, W. 1938b. Gestalt und Größe von *Brachiosaurus* und anderen riesenwüchsigen Sauropoden. Der Biologe, 7: 130–134.
Linnaeus, C. 1758. Systema naturae per regna tria naturae, secundum classes, ordines, genera, species, cum characteribus, differentiis, synonymis, locis. Tomus I. Editio decima, reformata. 503 pp. Stockholm (Salvius).
Makarieva, A. M., V.G. Gorshkov & B.-L. Li. 2005. Temperature-associated upper limits to body size in terrestrial poikilotherms. Oikos, 111: 425–436.
Mallison, H. 2010a. The digital *Plateosaurus* I: Body mass, mass distribution and posture assessed using CAD and CAE on a digitally mounted complete skeleton. Palaeontologia Electronica 13, (2; 8A): 26p; http://palaeo-electronica.org/2010_2/198/index.html
Mallison, H. 2010b. The digital *Plateosaurus* II: An assessment of the range of motion of the limbs and vertebral column and of previous reconstructions using a digital skeletal mount. Acta Palaeontologica Polonica 55(3): 433–458. doi:10.4202/app.2009.0075
Mallison, H. 2011. Defense capabilites of *Kentrosaurus aethiopicus* Hennig, 1915. Palaeontologia Electronica. in press.
Marsh, O.C. 1877a. Notice of new dinosaurian reptiles from the Jurassic formation. American Journal of Science and Arts 14: 514–516.
Marsh, O.C. 1877b. A new order of extinct Reptilia (Stegosauria) from the Jurassic of the Rocky Mountains. American Journal of Science 3(14): 513–514.
Marsh O.C. 1878. Principal characters of American Jurassic dinosaurs. Part I. American Journal of Science, 3: 411–416.
Mazzetta, G.V., Christiansen, P., & Fariña, R.A. 2004. Giants and bizarres: body size of some Southern South American Cretaceous dinosaurs. Historical Biology, 16: 71–83
Meyer, H.v. 1837. Mitteilung an Prof. Bronn (Message to Prof. Bronn). Neues Jahrbuch für Geologie und Paläontologie, 316.

Monbaron, M, DA Russell & P Taquet 1999: Atlasaurus imelakei n.g., n.sp., a brachiosaurid-like sauropod from the Middle Jurassic of Morocco. Comptes Rendus de l'Academie des Sciences ter. plan, 329: 519–526.

Paul, G.S. 1987. The science and art of restoring the life appearance of dinosaurs and their relatives: a rigorous how-to guide. Pp. 5–49. In Czerkas, S.J. and Olson, E.C. (eds.), Dinosauria Past and Present Vol. II. University of Washington Press, Seattle.

Paul, G.S. 1997. Dinosaur models: the good, the bad, and using them to estimate the mass of dinosaurs. Pp. 129–154. In Wolberg, D.L., Stump, E., and Rosenberg, G. (eds.), Dinofest International: Proceedings of a Symposium held at Arizona State University.

Paul, G.S. 2000. Restoring the life appearance of dinosaurs. Pp. 78–106. In Paul, G.S. (ed.), The Scientific American Book of Dinosaurs. Byron Press and Scientific American, New York

Perry, S. F. 2001. Functional morphology of the reptilian and avian respiratory systems and its implications for theropod dinosaurs. In New perspectives on the origin and early evolution of birds. J. Gauthier and L. F. Gall (eds.). Yale Peabody Museum, New Haven, Conn, pp. 429–441.

Persons, W. 2009. Theropod tail muscle reconstruction and assessment of the locomotive contributions of the M. caudofemoralis. Journal of Vertebrate Paleontology, 29(Supplement to No.3):164A.

Peters, R. H. 1983. The Ecological Implications of Body Size. Cambridge University Press, New York, 329 pp.

Riggs, E.S. 1903. *Brachiosaurus altithorax*, the largest known dinosaur. American Journal of Science, 4(15): 299–306.

Sander, P.M. 1992. The Norian Plateosaurus bonebeds of central Europe and their taphonomy. Palaeogeography, Palaeoclimatology, Palaeoecology, 93(3–4): 255–299.

Sander, P.M., Christian, A., Clauss, M., Fechner, R., Gee, C., Griebeler, E.-M., Gunga, H.-C., Hummel, J., Mallison, H., Perry, S., Preuschoft, H., Rauhut, O., Remes, K., Tütken, T., Wings, O., and Witzel, U. 2010. Biology of the Sauropod Dinosaurs: The Evolution of Gigantism. *Biology Letters*, 68(1): 117–155; doi:10.1111/j.1469-185X.2010.00137.x

Schmidt-Nielsen, K. 1984. Scaling: why is animal size so important? Cambridge University Press, Cambridge, 241 pp.

Schmidt-Nielsen, K. 1997. Animal physiology: adaptation and environment. 5th ed. Cambridge University Press, Cambridge, ix + 607 pp.

Schwarz, D. & G. Fritsch. 2006. Pneumatic structures in the cervical vertebrae of the Late Jurassic (Kimmeridgian-Tithonian) Tendaguru sauropods *Brachiosaurus brancai* and *Dicraeosaurus*. Eclogae Geologicae Helvetiae, 99: 65–78.

Schwarz-Wings, D. 2009. Approach to the reconstruction of the thoracic epaxial musculature of sauropod dinosaurs. – Journal of Vertebrate Paleontology, 29(2): 517–534

Schwarz-Wings, D., E. Frey, and C.A. Meyer. 2007. Pneumaticity and soft-tissue reconstructions in the neck of diplodocid and dicreaosaurid sauropods. Acta Palaeontologica Polonica, 52: 167–188.

Seebacher, F. 2001. A new method to calculate allometric length-mass relationships of dinosaurs. Journal of Vertebrate Paleontology 21, 51–60.

Taylor, M.P., 2009. A re-evaluation of Brachiosaurus altithorax Riggs 1903 (Dinosauria, Sauropoda) and its generic separation from Giraffatitan brancai (Janensch 1914). Journal of Vertebrate Paleontology, 29: 787–806.

Therrien, F. & Henderson, D.M. 2007. My theropod is bigger than yours— or not: Estimating body size from skull length in theropods. Jounral of Vertebrate Paleontology, 27:108–115

Wedel, M.J. 2003a. Vertebral pneumaticity, air sacs, and the physiology of sauropod dinosaurs. Paleobiology, 29: 243–255.

Wedel, M.J. 2003b. The evolution of vertebral pneumaticity in sauropod dinosaurs. Journal of Vertebrate Paleontology, 23: 344–357.

Wedel, M.J. 2005. Postcranial skeletal pneumaticity in sauropods and its implications for mass estimates. In The Sauropods: Evolution and Paleobiology. Wilson, J.A., and Curry-Rogers, K. (eds.). University of California Press, Berkeley, pp. 201–228.

Wiedemann, Albert, Tim Suthau & Jörg Albertz 1999. Photogrammetric Survey of Dinosaur Skeletons. Mitteilungen des Museums für Naturkunde Berlin, Geowissenschaftliche Reihe, 2: 113–119.

Young, C.C. 1941. A complete osteology of *Lufengosaurus huenei* Young (gen. et sp. nov.) from Lufeng, Yunnan, China. Palaeontologica Sinica, Series C, 7:1–53.

Young, C.C., & Zhao, X.-J. 1972. *Mamenchisaurus hochuanensis* sp. nov. Institute of Vertebrate Paleontology and Paleoanthropology Monographs, A, 8:1–30.

Zhang, Yihong; Li, kui; Zeng, Qinghua 1998. A new species of sauropod from the Late Jurassic of the Sichuan Basin (*Mamenchisaurus jingyanensis* sp. nov.). Journal of the Chengdu University of Technology, 25 (1): 61–68.

Chapter 9
Rates of Cranial Evolution in Neandertals and Modern Humans

Timothy D. Weaver

Introduction

Rates of phenotypic evolution have been a persistent interest of paleontologists (e.g., Eldredge and Gould 1972; Gingerich 1983; Hunt 2007; Stanley 1985), because of their potential to illuminate the mechanisms that generate or constrain phenotypic diversification within and between evolutionary lineages over long time scales. Field or laboratory studies of phenotypic evolution conducted generation by generation can provide detailed insights into evolutionary mechanisms (e.g., Grant and Grant 2002), but they may give an unrepresentative picture of the relative importance of different evolutionary processes over long time scales, because they typically document, at most, tens of generations. It is possible, perhaps probable, that the tens of generations that were observed in the field or laboratory are unrepresentative of the tens of thousands or hundreds of thousands of generations of existence of a particular set of taxa. In contrast, phenotypic comparisons between extant taxa whose evolutionary lineages are known to have split from each other many generations in the past, between fossil and extant taxa, or between different fossil taxa can document evolutionary diversification over long time scales, albeit with a much coarser resolution than field or laboratory studies. As such, both kinds of studies provide complementary information about evolutionary pattern and process.

When investigating rates of phenotypic evolution from either extant or fossil taxa, a neutral model of phenotypic evolution provides a useful baseline against which empirically estimated rates can be compared (Lande 1976; Lynch 1990). Under complete neutrality (i.e., natural selection is not acting at all), genetic drift provides the mechanism, and mutation provides the raw material for evolutionary change (Kimura 1968, 1989; Lynch 1990). Evolutionary rates faster than expected under complete neutrality could indicate that unidirectional natural selection has dominated over the time slice under consideration. Unidirectional natural selection

T.D. Weaver
Department of Anthropology, University of California, One Shields Avenue, Davis, CA 95616, USA
e-mail: tdweaver@ucdavis.edu

A.M.T. Elewa (ed.), *Computational Paleontology*,
DOI 10.1007/978-3-642-16271-8_9,

occurs when individuals with trait values that deviate from the species mean in a particular direction (e.g., larger than the mean) have higher reproductive success than other members of the species consistently over many generations. Conversely, slower rates could reflect strong stabilizing natural selection or fluctuating natural selection, because both kinds of natural selection tend to maintain the status quo over long periods of time. With stabilizing natural selection, individuals with trait values close to the mean consistently have the highest reproductive success, and with fluctuating natural selection, high reproductive success alternates between individuals with larger trait values and those with smaller trait values. When multiple traits are considered at once, the fit with neutral expectations can be evaluated by averaging over all traits and all groups (populations or species) or by considering distributions across traits or groups.

Here I investigate the rates of phenotypic evolution in Neandertals (*Homo neanderthalensis*) and modern humans (*Homo sapiens*). I focus on cranial evolution and make reference to both neutral expectations and the rates of cranial evolution in chimpanzees. The grouping "modern humans" includes present-day humans and fossils that are widely thought to document the evolutionary lineage leading to present-day humans, as separate from the lineages leading to Neandertals and other extinct human taxa. Neandertals are of particular interest in human paleontology because, based on current evidence, their evolutionary lineage is the one that split most recently from the lineage leading to present-day humans. Molecular and fossil evidence suggest that this split occurred ~350,000 years ago (Green et al. 2010; Noonan et al. 2006; Stringer and Hublin 1999; Weaver et al. 2008) (Fig. 9.1). Fossils that are classified as Neandertals in a narrow sense are present in the fossil record by ~130,000 years ago (Hublin 1998; Rink et al. 1995) and persist until <35,000 years ago (Higham et al. 2006; Hublin et al. 1995).

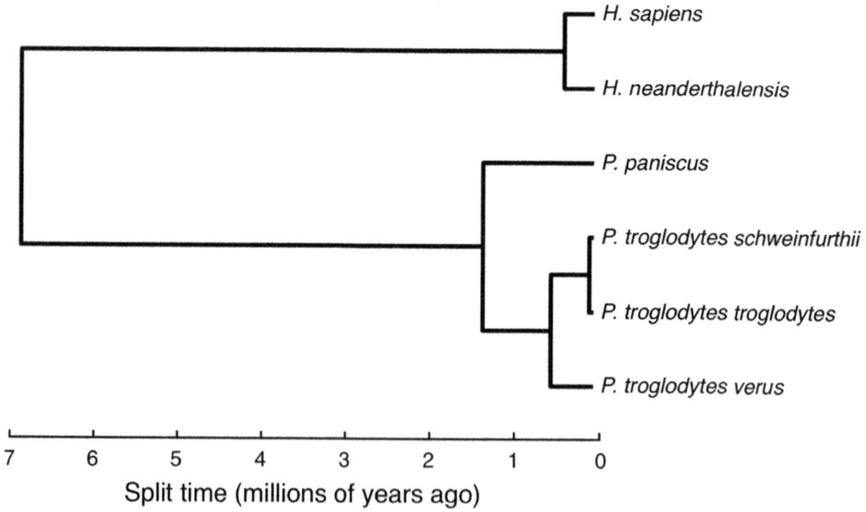

Fig. 9.1 A working phylogeny of modern humans, Neandertals, and chimpanzees. Based on Caswell et al. (2008), Green et al. (2010), Noonan et al. (2006), Stringer and Hublin (1999), and Weaver et al. (2008)

The split between the Neandertal and modern human evolutionary lineages may have been precipitated by geographic barriers produced by climate fluctuations that isolated Neandertal populations in Europe from modern human populations further south (Howell 1952; Hublin 1998). The ancestors of modern humans originated in Africa and began to populate the rest of the world ~50,000 years ago (Fleagle and Gilbert 2008; Klein 2009; and accompanying articles in these special journal issues on modern human origins).

Cranial Divergence Between Neandertals and Modern Humans

The crania of Neandertals typically differ from those of modern humans in numerous features (Franciscus 2002, 2003; Harvati 2003, 2004; Holton and Franciscus 2008; Spoor et al. 2003; Trinkaus 2006; Weaver 2009) (Fig. 9.2).

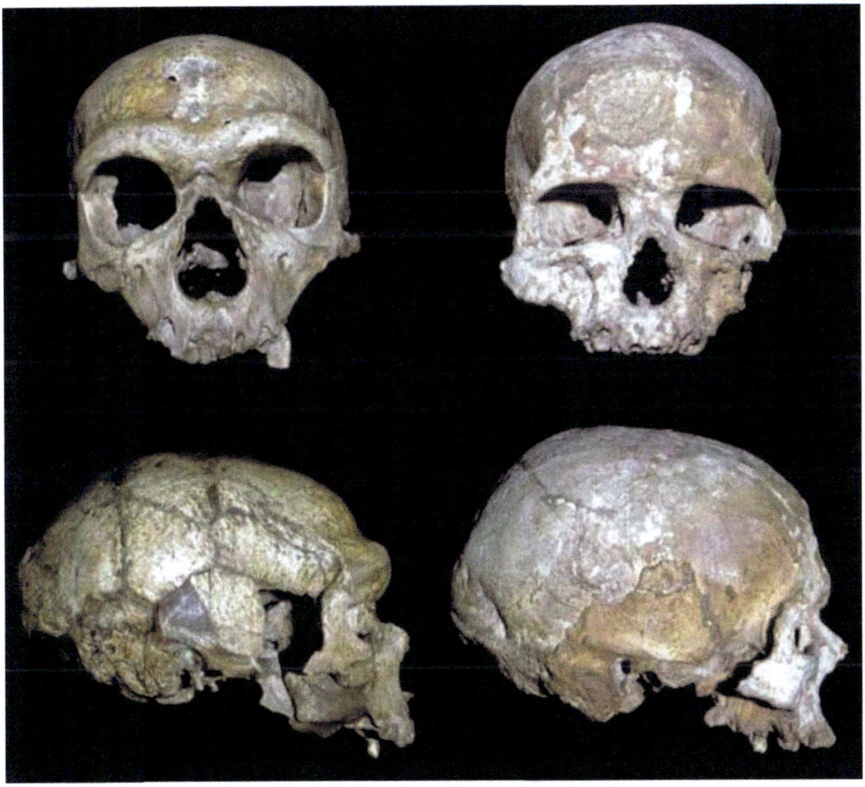

Fig. 9.2 Neandertal and modern human cranial differences. On the *left* is a Neandertal (La Chapelle-aux-Saints). On the *right* is an early modern human fossil (Cro-Magnon 1). Front (*above*) and side (*below*) views. Photos courtesy of Chris Stringer and the Musée de l'Homme (Paris)

Table 9.1 Selected features of the Neandertal cranium

Cranial vault
Receding frontal squama
Long, low braincase, sometimes with a posteriorly bulging occipital ("bun")
Globular ("en-bombe") braincase when viewed from behind
Occipital torus with a suprainiac fossa above it
Cranial base
Fairly unflexed ectobasicranium
Large juxtamastoid eminence and relatively small mastoid process
Tubercle on mastoid process adjacent to the external auditory meatus
Anteroposteriorly elongated foramen magnum
Acoustic cavities
Large lateral, small anterior, and small and inferiorly positioned posterior semicircular canal
Facial skeleton
Pronounced, double-arched supraorbital torus
Projecting midface
Minimally angled zygomatic bone
Absence of infraorbital concavity and canine fossa
Wide, tall nasal aperture
Wide, projecting nasal bridge
Depressed nasal floor

In comparison with modern humans, Neandertals have receding foreheads, accentuated by large brow ridges, that connect to long, low braincases with rounded sides when viewed from behind. Their faces are dominated by wide, tall, and projecting noses and jutting jaws with large front teeth. Unlike the midline of the face, the sides of the Neandertal face are not especially projecting, which leads to a smooth and swept-back appearance of the cheek region. In addition to these large-scale contrasts, Neandertals differ from modern humans in many detailed features (Table 9.1).

One way to compare the rates of cranial evolution along the Neandertal and modern human evolutionary lineages with neutral expectations is to use cranial differences between the two groups to estimate when the two lineages split and compare these results with those based on ancient Neandertal and present-day human DNA sequences. If the rates of cranial divergence are faster than neutral expectations, then split time estimates based on cranial differences should be older than those based on DNA sequences. Conversely, slower phenotypic divergence should lead to cranial estimates of split time that are too recent. Importantly, because information about the cranial anatomy of the last common ancestor of Neandertals and modern human is not incorporated into the comparison, it is not possible with this approach to determine if the rates were similar along both lineages or if one lineage evolved more rapidly than the other one.

My colleagues and I performed such a comparison based on 37 standard cranial measurements collected on 2,524 present-day humans from 30 globally distributed populations and 20 Neandertal specimens (Weaver et al. 2008). The measurements on the 2,524 present-day humans are a subset of Howells' craniometric dataset (Howells 1973, 1989, 1995). We adapted a split time estimator that was originally developed for molecular variation in the number of short tandem repeats (STRs) for use with variation in cranial measurements. We used STR estimates of the split time between sub-Saharan African and other present-day human populations and of the effective population size (number of breeding individuals in an idealized population that would have as much genetic drift as the actual population) of sub-Saharan African populations to calibrate the split time estimator. Specifically, we equated the amount of cranial variation between sub-Saharan African and other present-day human populations and within sub-Saharan African populations with the STR estimates (Zhivotovsky et al. 2003) of split time and effective population size respectively. This calibration implicitly assumes that patterns of cranial variation within and among modern human populations are the result of neutral evolution (see section "Cranial Divergence Among Present-Day Human Populations"). If this assumption of neutrality were false, then a close correspondence between cranial and molecular estimates of the Neandertal and modern human split time would simply indicate that the rates of cranial evolution along the Neandertal and modern human evolutionary lineages were similar to those that led to the cranial diversity of present-day human populations. In other words, both Neandertals and modern humans diverged in cranial form from the last common ancestor of Neandertals and modern humans at the same rate as present-day human populations diverged in cranial form from the last common ancestor of present-day human populations. Using the cranial split time estimator, we calculated that the evolutionary lineages leading to Neandertals and modern human split ~311,000 or ~435,000 years ago, depending on assumptions about changes in within-group variation. Consistent with neutral expectations, these dates are quite similar to those derived from ancient Neandertal and present-day human DNA sequences (Noonan et al. 2006; Weaver et al. 2008).

In principle, instead of averaging over all 37 measurements, each cranial measurement individually could be used to estimate a split time and the distribution of split time estimates across different measurements could be compared with neutral expectations. In practice, because the cranium is an integrated structure (i.e., cranial measurements vary together because of genetic, developmental, or functional links), the raw cranial measurements cannot be used directly to produce a split time distribution. To factor out the covariance among measurements produced by integration, my colleagues and I created a new set of 37 measurements as the eigenvectors (principal components) of the pooled within-population variance-covariance matrix for the present-day human populations (Weaver et al. 2007). Then, for each eigenvector we calculated the ratio of the variance between Neandertals and modern humans to the variance within present-day human populations (eigenvalues). Each of these cranial distances can be thought of as unscaled split time estimates, and together they form a split time distribution. Once we had created a split time distribution for Neandertals and modern humans, we compared it with neutral expectations using three statistical tests, based on the variance, shape, and patterning of the distribution. We were unable to reject neutrality at the $\alpha = 0.05$ level with any of the statistical tests. Based on current evidence, patterns of cranial divergence between Neandertals and modern humans fit closely with neutral expectations.

Deviations from neutral divergence along either the Neandertal or the modern human evolutionary lineages, or both, as well as shifts in the within-population variance–covariance matrix, could, in principle, all lead to rejection (Weaver et al. 2007), but more work is needed to determine the sensitivity of the statistical tests to different kinds of deviations. Additionally, the 37 measurements my colleagues and I considered in our two studies (Weaver et al. 2007, 2008) only reflect aspects of cranial form that can be readily quantified with standard osteometric tools, and the measurements sometimes span different cranial regions (e.g., from the base to the face), so a somewhat different picture might emerge from detailed studies of individual cranial regions.

Cranial Divergence Among Present-Day Human Populations

Rates of cranial evolution along just the modern human evolutionary lineage, as opposed to lumping the Neandertal and modern human lineages together, can be investigated with comparisons of present-day human populations. Because present-day human populations split from each other at most 150,000–50,000 years ago (Zhivotovsky et al. 2003), these comparisons reflect a fraction of the ~350,000 years since the modern human evolutionary lineage split from the Neandertal lineage (Green et al. 2010; Noonan et al. 2006; Stringer and Hublin 1999; Weaver et al. 2008), but they still provide insight into the processes underlying modern human cranial evolution. Investigating cranial evolution along just the modern human evolutionary lineage is important, because it may have proceeded quite differently from what happened along the Neandertal lineage.

Present-day human populations differ, on average, in certain features of the cranium. Many of these features are found in the face, particularly the nose, but there are also differences in other cranial regions (Byers 2002; Gill 1998; Hennessy and Stringer 2002; Howells 1973, 1989, 1995). For example, Europeans tend to have narrow, tall, and projecting noses, whereas sub-Saharan Africans tend to have wide, short, and flat noses. When one is interested in rates of cranial evolution in modern humans the basic question is: how different are the crania of present-day human populations from each other with respect to neutral expectations?

Both Lynch (1989) and Relethford (1994) compared present-day human cranial variation, as quantified by subsets of Howells' craniometric dataset (Howells 1973, 1989, 1995), to neutral expectations with approaches that average over multiple cranial measurements and human populations. Lynch (1989) performed a number of analyses, but, of particular relevance here, he evaluated whether the mutation rate (average amount of new additive genetic variance introduced by mutation per zygote per generation) required to produce cranial differences among human populations was similar to mutation rates estimated from experimental studies of a number of different taxa for a variety of traits. This approach is comparable to the one taken by my colleagues and I (Weaver et al. 2008), except that instead of assuming a mutation rate and calculating a split time, Lynch (1989) assumed a split time and calculated a mutation rate. For neutral evolution, the mutation rate largely determines the divergence rate (Lynch and Hill 1986), unless there are fluctuations in effective population size (Weaver et al. 2008), so correspondence between mutation rates calculated from divergence data and from experiments indicates consistency with neutral expectations. In line with neutral expectations, Lynch (1989) calculated a mutation rate based on present-day human cranial divergence within the range expected from experimental studies.

Instead of focusing on split times, which may not be an appropriate model within a species, because of the potential for reticulation (gene flow between populations), Relethford (1994) compared F_{ST} (a measure of among-population differentiation) estimates from cranial measurements to those from presumably neutral genetic loci. Like Lynch (1989), Relethford (1994) based his analyses on a subset of Howells' craniometric dataset (Howells 1973, 1989, 1995). In both the cranial and genetic case, F_{ST} measures differentiation at the level of genome, either directly or, in the case of the cranial measurements, indirectly. Consequently, if the two F_{ST} estimates are similar, then the rate of cranial divergence among present-day human populations is consistent with neutral expectations. Relethford (1994, 2002) found that cranial estimates of F_{ST} range from 0.07 to 0.15, depending on which human populations are considered and assumptions about heritability, which is in line with F_{ST} estimates for presumably neutral genetic loci between about 0.05 and 0.15 (Brown and Armelagos 2001; Relethford 1994, 2002; Rosenberg et al. 2002).

To further test for deviations from neutral expectations, my colleagues and I (Weaver et al. 2007) compared the distributions of unscaled split time estimates for different aspects of cranial variation (see section "Cranial Divergence Between Neandertals and Modern Humans" for more details) to the distributions for three categories of STRs using statistical tests based on the variance and shape of the distributions. These analyses average over all populations, but because they are based on distribution statistics other than the mean, they, at least to a certain extent, consider different aspects of cranial variation separately. Consistent with neutral expectations, the mean values for variance and shape for cranial measurements, averaged over all pairwise comparisons of the 30 present-day human populations in Howells' craniometric dataset (Howells 1973, 1989, 1995), were similar to those for STRs, averaged over all pairwise comparisons of the populations in Rosenberg and colleagues' STR dataset (Rosenberg et al. 2002; Zhivotovsky et al. 2003). Essentially, based on the comparisons we made, cranial measurements look as neutral as STRs.

A number of studies have compared distance or affinity matrices for present-day human populations based on cranial measurements with analogous distance or affinity matrices based on presumably neutral genetic loci. These analyses average over all measurements, but they consider pairs of populations separately. Similar to the situation with F_{ST}, the neutral expectation is correspondence between the cranial and genetic distance matrices, as long as compatible distance statistics are used (i.e., define distance in an analogous way). Unlike the situation with F_{ST}, the cranial and genetic distances are not expected to be strictly equal, but they should be proportional. In line with neutral expectations, cranial and genetic distance matrices of the same or similar present-day human populations are significantly correlated with each other (Gonzáles-José et al. 2004; Harvati and Weaver 2006a, b; Roseman 2004; Smith 2009; von Cramon-Taubadel 2009). Importantly, unlike the studies by Lynch (1989), Relethford (1994), and my colleagues and I (Weaver et al. 2007) that are all based on the same craniometric dataset, these distance matrix comparisons are based on four different craniometric datasets.

Cranial distance matrices have also been compared to geographic distance matrices. Geographic distance, particularly when calculated with waypoints around large bodies of water, is an excellent predictor of genetic distance for worldwide datasets of present-day human populations (Ramachandran et al. 2005; Relethford 2004). Consequently, as with genetic and cranial distance matrices, the neutral expectation is proportionality of geographic and cranial distance matrices. In line with neutral expectations, cranial and geographic distance matrices are significantly correlated with each other (Betti et al. 2009; Hubbe et al. 2009; Relethford 2004). Two of these studies are based on Hanihara's craniometric dataset (Hanihara 2008; Manica et al. 2007) and one of them is based on Howells' dataset (Howells 1973, 1989, 1995), so there is at least some consistency in results across different datasets.

The general picture that emerges from studies of present-day human cranial diversity is that the rates of phenotypic evolution are consistent with neutral expectations (Roseman and Weaver 2007; von Cramon-Taubadel and Weaver 2009). This is not to say that certain measurements or populations never show deviations from neutral expectations. For example, F_{ST} estimates are higher for certain nasal measurements, suggesting that different human populations have experienced divergent directional selection on nose size and shape (Roseman and Weaver 2004). This result is perhaps not surprising given that the nose plays an important role in heat and moisture transfer, and that multiple studies have documented significant relationships between human nasal form and climatic variables (Carey and Steegmann 1981; Franciscus and Long 1991). Additionally, deviations from neutrality appear to be more pronounced in analyses that include human populations from very high latitudes (Betti et al. 2009; Harvati and Weaver 2006a; Hubbe et al. 2009; Roseman 2004; von Cramon-Taubadel 2009; von Cramon-Taubadel and Weaver 2009).

Comparisons with Chimpanzees

Chimpanzees (*Pan troglodytes* and *Pan paniscus*) are the extant species most closely related to humans. As such, they provide a broader comparative context in which to evaluate the rates of cranial evolution in Neandertals and modern humans. Of particular interest are comparisons of the western (*Pan troglodytes verus*) subspecies of common chimpanzee with the central (*Pan troglodytes troglodytes*) and eastern (*Pan troglodytes schweinfurthii*) subspecies (Fig. 9.3).

The western subspecies split from other common chimpanzees ~510,000 years ago (Caswell et al. 2008), which is similar to the split time between Neandertals and

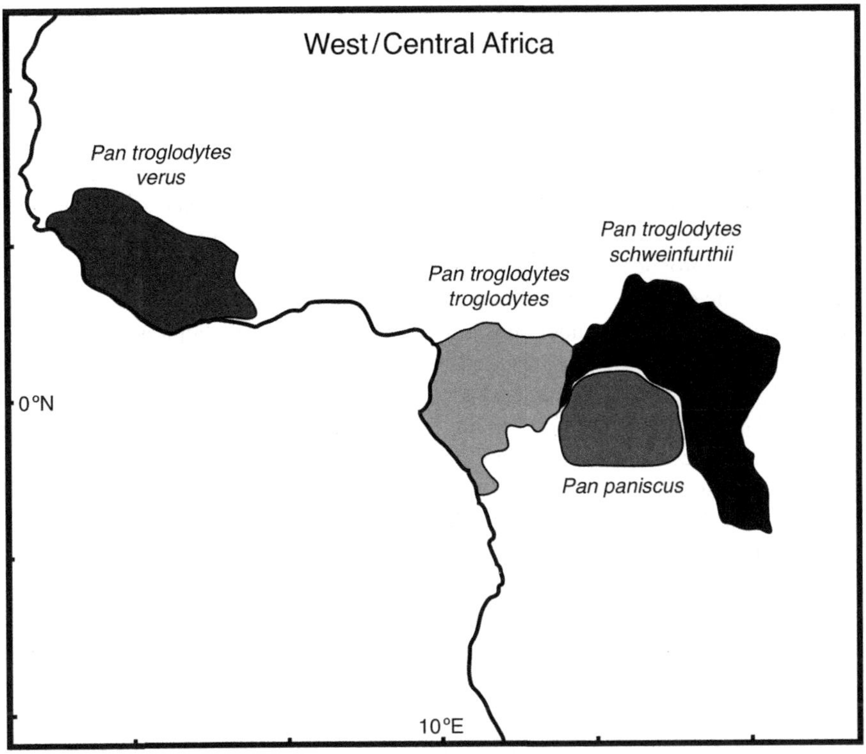

Fig. 9.3 Map of West/Central Africa showing the geographic ranges of the chimpanzee species and subspecies discussed in the text. Based on Caldecott and Kapos (2005) and Inskipp (2005)

modern humans of ~350,000 years ago (Green et al. 2010; Noonan et al. 2006; Stringer and Hublin 1999; Weaver et al. 2008) (Fig. 9.1). Across a number of different taxa for a variety of traits there is typically an inverse relationship between the rate of phenotypic evolution and split time (Gingerich 1983, 2001, 2009; Kinnison and Hendry 2001; Lynch 1990; Roopnarine 2003). For example, if taxa A1 and A2 split from each other twice as long ago as taxa B1 and B2 split from each other, then the amount of morphological divergence between taxa A1 and A2 will tend to be less than twice as much as the amount of morphological divergence between taxa B1 and B2. There are a number of candidate explanations for the time dependency of phenotypic rates (see discussion in Roopnarine 2003), but regardless of the reason for this empirical observation, the upshot is that it is difficult to infer anything about evolutionary process from rate comparisons unless the timescales are similar. Hence, comparisons of Neandertals and modern humans with common chimpanzees subspecies are worthwhile for both phylogenetic and a temporal (i.e., split time) reasons.

Unfortunately, researchers studying cranial variation in Neandertals and/or present-day human populations have tended to not include non-human primates in their samples, and the converse has been true for researchers investigating non-human primate cranial variation. One exception is the study by Harvati and colleagues of cranial variation in Neandertals, modern humans, African apes, and Old World Monkeys (Harvati et al. 2004). These researchers collected 3D coordinate data on 15 standard anatomical landmarks of the cranium. Importantly, they selected landmarks that summarized overall cranial shape, rather than picking landmarks on features known to be particularly good at distinguishing Neandertals from modern humans, present-day human populations from each other, or non-human primate species or subspecies from each other, so the results are not biased to necessarily produce larger or smaller cranial distances for certain taxonomic groupings. The distance measure that they used, Mahalanobis D^2, is expected to be proportional to split time under neutrality (Lynch 1990). Mahalanobis D^2 measures rates of phenotypic evolution averaged over all measurements (landmarks). Harvati and colleagues (2004) calculated mean $D^2 = 109.7$ between Neandertals and modern humans, averaged over all present-day and fossil human groups; $D^2 = 43.2$ between the western and central common chimpanzee subspecies; and $D^2 = 38.2$ between the western and eastern common chimpanzee subspecies. Even though Neandertals split from modern humans approximately as far back in time as the western common chimpanzee subspecies split from other chimpanzees, the amount of cranial divergence between Neandertals and modern humans is more than twice as large. In fact, the mean $D^2 = 43.7$ among present-day human populations is most similar to the common chimpanzee subspecies comparisons, even though present-day human populations split from each other at most 150,000–50,000 years ago (Zhivotovsky et al. 2003). The bottom line is that the rates of cranial evolution appear to be markedly faster in Neandertals and modern humans than in common chimpanzees.

Explanations for Accelerated Rates in Neandertals and Modern Humans

There are at least three possible explanations for the accelerated rates of cranial evolution in Neandertals and modern humans relative to common chimpanzees. It may be that stabilizing natural selection has affected the common chimpanzee cranium more than the Neandertal and modern human cranium. As I discussed above, stabilizing natural selection will tend to slow down rates of phenotypic evolution. Why stabilizing selection would be more prevalent in common chimpanzees than in Neandertals and modern humans is not obvious. One possibility is that culture (i.e., technology) has buffered Neandertals and modern humans, at least to a certain extent, from stabilizing selection. A second possibility is that stabilizing selection has been weaker in Neandertals and modern humans because of a smaller effective population size than in common chimpanzees. Relative to genetic drift, natural selection is weaker in populations with small effective sizes (Ohta 1996), which will tend to lead to more rapid rates of evolution if the stabilizing natural selection in the predominant kind of natural selection. This explanation is consistent with the reduced genetic diversity, presumably due to small effective population sizes, for Neandertals and modern humans relative to common chimpanzees (Caswell et al. 2008; Gagneux et al. 1999; Green et al. 2008; Kaessmann et al. 2001; Krause et al. 2007). Finally, it may be that cranial evolution in Neandertals and modern humans is less genetically and/or developmentally constrained than in common chimpanzees, although it is difficult to imagine why would there be such different constraints in closely related species. Further study is needed to distinguish among these three candidate explanations.

Conclusions

Multiple comparisons between Neandertals and modern humans and among present-day humans' populations suggest that the rates of cranial evolution in these groups are consistent with neutral expectations. It appears that cranial divergence has followed a sort of "morphological clock" analogous to the "molecular clock", and, in general, the phenotype and genotype show similar patterns of variation. In contrast, based on the admittedly limited available data, the rates of cranial divergence among subspecies of common chimpanzees appear to have been considerably slower. The sum may indicate that cranial divergence in common chimpanzees has been constrained by stabilizing natural selection, while Neandertals and modern humans have been less restricted, leading to greater cranial diversity.

Acknowledgements I thank Ashraf Elewa for inviting me to contribute this chapter; Chris Stringer and the Musée de l'Homme (Paris) for the photos for Fig. 9.2; Andy Marshall for help with Fig. 9.3; and Philipp Gunz and Noreen von Cramon-Taubadel for comments.

References

Betti L, Balloux F, Hanihara T, Manica A (2009) The relative role of drift and selection in shaping the human skull. Am J Phys Anthropol 141:76–82

Brown RA, Armelagos GJ (2001) Apportionment of racial diversity: A review. Evolutionary Anthropology 10:34–40

Byers SN (2002) Introduction to forensic anthropology: A textbook. Allyn and Bacon, Boston

Caldecott J, Kapos V (2005) Great ape habitats: tropical moist forests of the Old World. In: Caldecott J, Miles L (eds) World atlas of great apes and conservation. University of California Press, Berkeley, pp 31–42

Carey JW, Steegmann AT (1981) Human nasal protrusion, latitude, and climate. Am J Phys Anthropol 56:313–319

Caswell J, Mallick S, Richter DJ, Neubauer J, Schirmer C, Gnerre S, Reich D (2008) Analysis of chimpanzee history based on genome sequence alignments. PLoS Genetics 4:e1000057

Eldredge N, Gould SJ (1972) Punctuated equilibria: an alternative to phyletic gradualism. In: Schopf TJM (ed) Models in paleobiology. Freeman Cooper, San Francisco, pp 82–115

Fleagle JG, Gilbert CC (2008) Modern human origins in Africa. Evolutionary Anthropology 17:1–2

Franciscus RG (2002) Neanderthals. In: Pagel M (ed) Encyclopedia of evolution. Oxford University Press, Oxford, pp 493–497

Franciscus RG (2003) Internal nasal floor configuration in *Homo* with special reference to the evolution of Neandertal facial form. J Hum Evol 44:701–729

Franciscus RG, Long JC (1991) Variation in human nasal height and breadth. Am J Phys Anthropol 85:419–427

Gagneux P, Wills C, Gerloff U, Tautz D, Morin PA, Boesch C, Fruth B, Hohmann G, Ryder OA, Woodruff DS (1999) Mitochondrial sequences show diverse evolutionary histories of African hominoids. Proc Natl Acad Sci U S A 96:5077–5082

Gill GW (1998) Craniofacial criteria in the skeletal attribution of race. In: Reichs KJ (ed) Advances in the identification of human remains, 2nd edn. Charles C. Thomas, Springfield, pp 293–317

Gingerich PD (1983) Rates of evolution: effects of time and temporal scaling. Science 222:159–161

Gingerich PD (2001) Rates of evolution on the time scale of the evolutionary process. Genetica 112–113:127–144

Gingerich PD (2009) Rates of evolution. Annual Review of Ecology and Systematics 40:657–675

Gonzáles-José R, Van der Molen S, Gonzales-Perez E, Hernandez M (2004) Patterns of phenotypic covariation and correlation in modern humans as viewed from morphological integration. Am J Phys Anthropol 123:69–77

Grant PR, Grant BR (2002) Unpredictable evolution in a 30-year study of Darwin's finches. Science 296:707–711

Green RE, Malaspinas A-S, Krause J, Briggs AW, Johnson PLF, Uhler C, Meyer M, Good JM, Maricic T, Stenzel U, Prüfer K, Siebauer M, Burano HA, Ronan MT, Rothberg JM, Egholm M, Pavao R, Brajkovic D, Kucan Z, Gusic I, Wikström M, Laakkonen L, Kelso J, Slatkin M, Pääbo S (2008) A complete Neandertal mitochondrial genome sequence determined by high-throughput sequencing. Cell 134:416–426

Green RE, Krause J, Briggs AW, Maricic T, Stenzel U, Kircher M, Patterson N, Li H, Zhai W, Fritz MH-Y, Hansen NF, Durand EY, Malaspinas A-S, Jensen JD, Marques-Bonet T, Alkan C, Prüfer K, Meyer M, Burbano HA, Good JM, Schultz R, Aximu-Petri A, Butthof A, Höber B, Höffner B, Siegemund M, Weihmann A, Nusbaum C, Lander ES, Russ C, Novod N, Affourtit J, Egholm M, Verna C, Rudan P, Brajkovic D, Kucan Z, Gusic I, Doronichev VB, Golovanova LV, Laluenza-Fox C, de la Rasilla M, Fortea J, Rosas A, Schmitz RW, Johnson PLF, Eichler EE, Falush D, Birney E, Mullikin JC, Slatkin M, Nielsen R, Kelso J, Lachmann M, Reich D, Pääbo S (2010) A draft sequence of the Neanderthal genome. Science 328:710–722

Hanihara T (2008) Morphological variation of major human populations based on nonmetric dental traits. Am J Phys Anthropol 136:169–182

Harvati K (2003) Quantitative analysis of Neanderthal temporal bone morphology using three-dimensional geometric morphometrics. Am J Phys Anthropol 120:323–338

Harvati K (2004) 3-D geometric morphometric analysis of temporal bone landmarks in Neanderthals and modern humans. In: Elewa AMT (ed) Morphometrics: applications in biology and paleontology. Springer-Verlag, Heidelberg, pp 245–258

Harvati K, Weaver TD (2006a) Human cranial anatomy and the differential preservation of population history and climate signatures. The Anatomical Record 288A:1225–1233

Harvati K, Weaver TD (2006b) Reliability of cranial morphology in reconstructing Neanderthal phylogeny. In: Harvati K, Harrison T (eds) Neanderthals revisited: new approaches and perspectives. Springer, New York, pp 239–254

Harvati K, Frost SR, McNulty KP (2004) Neanderthal taxonomy reconsidered: implications of 3D primate models of intra- and interspecific differences. Proc Natl Acad Sci U S A 101(5): 1147–1152

Hennessy RJ, Stringer CB (2002) Geometric morphometric study of the regional variation of modern human craniofacial form. Am J Phys Anthropol 117(1):37–48

Higham T, Bronk Ramsey C, Karavanic I, Smith FH, Trinkaus E (2006) Revised direct radiocarbon dating of the Vindija G_1 Upper Paleolithic Neandertals. Proc Natl Acad Sci U S A 103(3): 553–557

Holton NE, Franciscus RG (2008) The paradox of a wide nasal aperture in cold-adapted Neandertals: a causal assessment. J Hum Evol 55:942–951

Howell FC (1952) Pleistocene glacial ecology and the evolution of "classic Neandertal man". Southwestern Journal of Anthropology 8(4):377–410

Howells WW (1973) Cranial variation in man: a study by multivariate analysis of patterns of difference among recent human populations. Harvard University, Cambridge

Howells WW (1989) Skull shapes and the map: craniometric analyses in the dispersion of modern *Homo*. Harvard University, Cambridge

Howells WW (1995) Who's who in skulls: ethnic identification of crania from measurements. Harvard University, Cambridge

Hubbe M, Hanihara T, Harvati K (2009) Climatic signatures in the morphological differentiation of worldwide human populations. The Anatomical Record 292:1720–1733

Hublin J-J (1998) Climatic changes, paleogeography, and the evolution of the Neandertals. In: Akazawa T, Aoki K, Bar-Yosef O (eds) Neandertals and modern humans in Western Asia. Plenum Press, New York, pp 295–310

Hublin J-J, Barroso Ruiz C, Medina Lara P, Fontugne M, Reyss J-L (1995) The Mousterian site of Zafarraya (Andalucia, Spain): dating and implications on the Palaeolithic peopling processes of Western Europe. Comptes Rendus de l'Academie des Sciences Paris 321:931–937

Hunt G (2007) The relative importance of directional change, random walks, and stasis in the evolution of fossil lineages. Proc Natl Acad Sci U S A 104:18404–18408

Inskipp T (2005) Chimpanzee (*Pan troglodytes*). In: Caldecott J, Miles L (eds) World atlas of great apes and conservation. University of California Press, Berkeley, pp 53–81

Kaessmann H, Wiebe V, Weiss G, Pääbo S (2001) Great ape DNA sequences reveal a reduced diversity and an expansion in humans. Nat Genet 27:155–156

Kimura M (1968) Evolutionary rate at the molecular level. Nature 217:624–626

Kimura M (1989) The neutral theory of molecular evolution and the world view of the neutralists. Genome 31:24–31

Kinnison MT, Hendry AP (2001) The pace of modern life II: from rates of contemporary microevolution to pattern and process. Genetica 112–113:145–164

Klein RG (2009) Darwin and the recent African origin of modern humans. Proc Natl Acad Sci U S A 106:16007–16009

Krause J, Orlando L, Serre D, Viola B, Prüfer K, Richards MP, Hublin J-J, Hänni C, Derevianko AP, Pääbo S (2007) Neanderthals in central Asia and Siberia. Nature 449:902–904

Lande R (1976) Natural selection and random genetic drift in phenotypic evolution. Evolution 30 (2):314–334

Lynch M (1989) Phylogenetic hypotheses under the assumption of neutral quantitative-genetic variation. Evolution 43(1):1–17

Lynch M (1990) The rate of morphological evolution in mammals from the standpoint of neutral expectation. American Naturalist 136(6):727–741

Lynch M, Hill WG (1986) Phenotypic evolution by neutral mutation. Evolution 40(5):915–935

Manica A, Amos W, Balloux F, Hanihara T (2007) The effect of ancient population bottlenecks on human phenotypic variation. Nature 448:346–349

Noonan JP, Coop G, Kudaravalli S, Smith D, Krause J, Alessi J, Chen F, Platt D, Pääbo S, Pritchard JK, Rubin EM (2006) Sequencing and analysis of Neanderthal genomic DNA. Science 314:1113–1118

Ohta T (1996) The current significance and standing of neutral and nearly neutral theories. Bioessays 18:673–677

Ramachandran S, Deshpande O, Roseman CC, Rosenberg NA, Feldman MW, Cavalli-Sforza LL (2005) Support from the relationship of genetic and geographic distance in human populations for a serial founder effect originating in Africa. Proc Natl Acad Sci U S A 102 (44):15942–15947

Relethford JH (1994) Craniometric variation among modern human populations. Am J Phys Anthropol 95:53–62

Relethford JH (2002) Apportionment of global human genetic diversity based on craniometrics and skin color. Am J Phys Anthropol 118:393–398

Relethford JH (2004) Global patterns of isolation by distance based on genetic and morphological data. Hum Biol 76:499–513

Rink WJ, Schwarcz HP, Smith FH, Radovcic J (1995) ESR ages for Krapina hominids. Nature 378:24

Roopnarine PD (2003) Analysis of rates of morphologic evolution. Annual Review of Ecology and Systematics 34:605–632

Roseman CC (2004) Detection of interregionally diversifying natural selection on modern human cranial form by using matched molecular and morphometric data. Proc Natl Acad Sci U S A 101(35):12824–12829

Roseman CC, Weaver TD (2004) Multivariate apportionment of global human craniometric diversity. Am J Phys Anthropol 125:257–263

Roseman CC, Weaver TD (2007) Molecules versus morphology? Not for the human cranium. Bioessays 29:1185–1188

Rosenberg NA, Pritchard JK, Weber JL, Cann HM, Kidd KK, Zhivotovsky LA, Feldman MW (2002) Genetic structure of human populations. Science 298:2381–2385

Smith HF (2009) Which cranial regions reflect molecular distances reliably in humans? Evidence from three-dimensional morphology. American Journal of Human Biology 21:36–47

Spoor F, Hublin J-J, Braun M, Zonneveld F (2003) The bony labyrinth of Neanderthals. J Hum Evol 44:141–165

Stanley SM (1985) Rates of evolution. Paleobiology 11:13–26

Stringer CB, Hublin J-J (1999) New age estimates for the Swanscombe hominid, and their significance for human evolution. J Hum Evol 37:873–877

Trinkaus E (2006) Modern human versus Neandertal evolutionary distinctiveness. Current Anthropology 47:597–620

von Cramon-Taubadel N (2009) Congruence of individual cranial bone morphology and neutral molecular affinity patterns in modern humans. Am J Phys Anthropol 140:205–215

von Cramon-Taubadel N, Weaver TD (2009) Insights from a quantitative genetic approach to human morphological evolution. Evolutionary Anthropology 18:237–240

Weaver TD (2009) The meaning of Neandertal skeletal morphology. Proc Natl Acad Sci U S A 106:16028–16033

Weaver TD, Roseman CC, Stringer CB (2007) Were Neandertal and modern human cranial differences produced by natural selection or genetic drift? J Hum Evol 53:135–145

Weaver TD, Roseman CC, Stringer CB (2008) Close correspondence between quantitative- and molecular-genetic divergence times for Neandertals and modern humans. Proc Natl Acad Sci U S A 105:4645–4649

Zhivotovsky LA, Rosenberg NA, Feldman MW (2003) Features of evolution and expansion of modern humans, inferred from genomewide microsatellite markers. Am J Hum Genet 72:1171–1186

Chapter 10
The Problem of Instar Numbers in Arthropods

Panos V. Petrakis

Introduction

The last 20 years the distribution of computing power to individuals through personal computers and the subsequent explosion in the production of scientific software dramatically changed the scientific work. The initial availability of expensive statistical software was replaced by inexpensive commercial packages – i.e., SPSS, SYSTAT, SAS, CART among others – which put in the hands of individuals with a little mathematical background and experience, the most powerful statistical procedures in the design of experimental projects, analysis of data and presentation of the results. Moreover the education of students and new scientists was facilitated by the advent of the educational (usually reduced in data-handling abilities and price software product) versions of the commercial packages and the free availability of specialized software. The PAST software program (Hammer et al. 2009) is one such a free package which incorporates a vast majority of analytical methods suitable for paleontological and current ecological work. Most of the methods in this package have been invented or refined in the last 40 years and are still in the research edge of scientific research. In the ecology, systematics and paleontology in particular, is very common for an introductory text book to reach at areas needing further research or modulation. Many cases and examples can be found in Legendre and Legendre (1983), Orloci and Kenkel (1985), Morrison (1984), Johnson and Wichern (1998), Hammer and Harper (2006).

Paleontology is greatly benefited from the explosion of electronic journals and other electronic publications. Apart from some problems associated with the bias of published electronic papers towards vertebrate paleontology, leaving uncovered the invertebrate and plant paleontology, and the credits given to electronic publications (Elewa 2007, 2009) it is expected that information technology in paleontology will solve many problems. For instance, the taxonomic problem of assigning two

P.V. Petrakis
National Agriculture Research Foundation-NAGREF, Institute of Mediterranean Forest Ecosystem Research, Laboratory of Entomology, Terma Alkmanos, 115 28 Ilissia, Athens, Greece
e-mail: pvpetrakis@fria.gr

A.M.T. Elewa (ed.), *Computational Paleontology*,
DOI 10.1007/978-3-642-16271-8_10,

different names to the same taxon will be solved if "paleoinformatics" is engaged in paleontological work. Also, the availability of many important paleontologically oriented databases will be available to researchers all over the world. Elewa (2010) suggests the mode of popularization of these databases by exploiting electronic journals and predicts that this is going to be the "superhighway to modern paleontology".

In biological systematics, ecology and paleontology the increased computing ability penetrated at data acquisition areas. Several acquisition analogue machines, photographic/scanning devices and bio-acoustic probes have extended their functionality accuracy and speed by exploiting the ability to digitize the probed signal and the related processing and analysis. For instance morphometric data collection (Hammer 2004) is much easier through the use of a three-dimensional digitizing apparatus that digitize the organism under consideration and depicts the acquired image on a computer screen. This morphometric collection set is then directly entered into statistical packages for processing and analysis.

Other data acquisition systems involve flatbed scanners commonly found on every desk as a part of the computer peripheral set (Bybee and Branham 2008). The method is routinely applied by taxonomists for the digitization of organisms ranging from extant insects to plant parts and fossils. Most scanners are capable to scan an area as small as a square with dimensions 2.5×2.5 mm with resolutions usually selected among 1,200; 2,400; 3,600 and 4,800 dots per inch (dpi) and pixel values at 48 bits (2^{48} colors or different wavelengths of the reflected radiation). The ability of multiple scans and the cheap storage space in the computer's hard disk makes the digitization work easy while the advent of DVDs and BlueRay® optical disks permitted the communication of scientists from different countries and the compilation of large image databases. Even advanced bio-mathematical work is possible by freeing the researcher from the data acquisition job. Such image processing packages are capable for particle counting, object classification, object tracking, densitometry among other more specialized facilities. Hunt and Chapman (2001) used these techniques to measure the cephalic dimensions of the trilobite *Chionaspis sellata* in a set of 38 individuals residing from a collection of the Field Museum of Natural History (FMNH PE 54116-54163). In their study they attempted to find a testable criterion for the detection of the number of instars in trilobites.

The available software today is vast. Apart from measurement or other data acquisition software there is a lot of programs to carry out general and more specialized analysis. For biologists specializing at organismal or ecological work there are general low-cost commercial packages. PAUP, PHYLIP and MacClade are two packages oriented to phylogeny, biological evolution, taxonomy and systematics. For paleontologists and biologists in other disciplines, the package PAST (Hammer et al. 2009) is a freely available package (freeware) not only suitable for students and larval researchers but also for more advanced scientific work. The setup of the R environment boosted the computational ability of desk top computing and greatly helped bio-statistical education (R Development Core Team 2008). Also the vast library associated with R includes a lot of packages, routines and functions suitable for any kind of ecological and paleontological work. The only requirement is that the users must have some familiarity with the – C like – language. R libraries associated with paleontology appeared in the library repository of the computing system (http://cran.r-project.org/src/contrib/Descriptions/). For paleoecology the package "analogue" (Simpson 2007) uses sub-fossilized remains of selected organisms to predict past environmental conditions of water bodies (lakes, oceans, rivers). In the package "adehabitat" presents the functions associated with the selection of habitat by animals. Although the paper examines only extant animals the applicability of methods can be extrapolated to extinct animals. Workers employing maximum likelihood methods for the estimation of model parameters are expected to benefit much from the "bio.infer" package (Yuan 2007). Apart from the computational facility of the package the author suggest many issues within the taxon–environment relationships. Other packages are listed in the respective parts of the text where they are used.

This work is a kind of journey to the facilities gained from the explosion of computing methods for a biostatistical treatment of a classical biological problem often encountered in the research of arthropod growth. This is the problem of the detection of the number of instars in arthropods, and is approached here through a statistical treatment of the various hypotheses.

The Problem of Instar Group Detection

Despite the continuous growth of body size of many organisms, invertebrates have a discontinuous stepwise growth in finite steps, usually called instars. This is caused partly because of the rigid exoskeleton that supports most invertebrates which is shed and a new exoskeleton is formed of bigger size to support the new increased body of the animal. Working with larvae of various Lepidoptera insect species Dyar (1890) estimated that the increase in size varied between 1.2 and 1.6 and in most cases was ~1.4. This value is taken as a reference by many authors although Hutchinson et al. (1997) found that the pattern described as Dyar's ratio is a special case where the proportional size increase after each molt (ecdysis = shedding of the old integument and bio-synthesize a new one) is constant. These authors used a growth equation (1) based on their Investment Principle to predict the optimal size increase after each molt, the optimal number of molting events (number of instars) and the adaptive significance of Dyar's ratio and the exceptions from this "Rule".

$$x_{i+1} = \varepsilon \chi_i + f \chi_i^{\alpha} d_i, \tag{1}$$

where x_i = size of the i-th instar; ε = efficiency of molting which equals the proportion of the mass of the old instar excluding those elements that can be accumulated to the new instar; f = constant related to the availability of food; d_i = duration of the i-th instar; α = rate of reserve accumulation. Under the general framework set with (1) and the other equations in the paper of Hutchinson et al. (1997) it is expected that the deviations from the expected optimal number of instars and the duration of instars can be caused by a multitude of constraints such as the tendency of some animals to molt a particular diel time, the availability of a restricted number of periods of seasonal time and ontogenetic noise imposed by the ecophysiology of the organism. Also of particular interest is that some insects molt but they do not increase in size while some species increase in size even within an instar stage.

In the majority of arthropods the increase in size is usually restricted to the larval instars. However, larval instars are not well preserved in short time periods since they have a few chitinized structures which are regularly exploitable in morphometrics. In insects the epicranial larval sclerites (or head capsules) are heavily exploited (Kishi 1971; Soponis and Russell 1982; Beaver and Sanderson 1989; Petrakis 2000; Goldson et al. 2001; Alvan-Aguilar and Hamada 2003; Lewandowski et al. 2004; Johnson and Williamson 2006). These structures are exploited also in other animal assemblages such as the extinct trilobites (Hunt and Chapman 2001). Many organisms have found other ways to cope with the restrictions imposed by the physical and biotic environment on the growth of immature stages. Polyphenism, that is the ability of adults to produce different adult morphs in response to environmental variation (Hanks and Denno 1993; Appleby and Credland 2007) is one adaptation of the organisms to change the reserves that the adult acquires from the environment. Even the larval behavior of some species can change in order to increase the resources that can be exploited; this fact, and some other, led Floater (1996) to split the Australian species *Ochrogaster lubifer* (Lepidoptera, Thaumetopoeidae) in two species.

Determination of the Number of Instars

The determination of the number of instars is the oldest question eventually in all morphometric analyses of extant insects (Daly 1985) and fossil arthropods (Kopaska-Merkel 1981). In cases where different larval instars have different behavior the alteration of living signs is exploited with fruitful results in extant species (e.g., Jiao et al. 1998; Petrakis 2000) but are rarely examined in paleoecological work. Paleontological work differs from current arthropod research in that in the former it is impossible to obtain titers of the biochemical compounds involved in the molting process such as the ecdysone and the regulators of molting glands (reviewed insects and crustaceans by Chang 1993). Since molting does not always results in increase in size the detection in titers of the compounds involved in the sclerotization of the newly formed integument (Hopkins and Kramer 1992) it is a much safer criterion which is also inapplicable in paleontology. However a property discovered by biochemists in extant insects and has not yet exploited in paleoentomology is that the type of cuticle sclerotization is reflected to the nature of involved compounds and the precursors of the respective biosynthetic pathways. The physical properties of the resulting sclerotized structure and in effect the fossilization processes may help for an indirect identification of "integument types" though the molting process is well study for some extinct groups (e.g., Whittington 1990).

In many studies the examination of the frequency distributions of a measurable trait is enough in detecting the discrete non-overlapping peaks. These peaks are assumed to correspond to one instar (Kishi 1971; Daly 1985). When several traits are simultaneously measured it might be that different peaks appear in the frequency diagrams of various traits measured as a linear size (Johnson and Williamson 2006). Since the organ size of various biological organisms is usually controlled by many genes the frequency distribution of the individuals of a population is usually normal. In this way the distinction of instars is problematic especially when different peaks correspond to normal distributions with overlapping lower parts (Fig. 10.1).

Another method based on Dyar's ratio [or Brooks-Dyar rule according to the proposal of Hutchinson and Tongring 1984 [in Hutchinson et al. (1997)] is based on the fact that Dyar's ratio must be consistent in all larval instars and as much as possible to the norm found for the animal assemblage. In this way is possible to find missing – i.e., no sampled – instars in one population. Many authors used this technique, sometimes reinforced with Crosby's ratio (1973) defined as Dyar's ratio of the current growth interval minus the Dyar's ratio of the previous growth interval and divided by the last Dyar's ratio, that is:

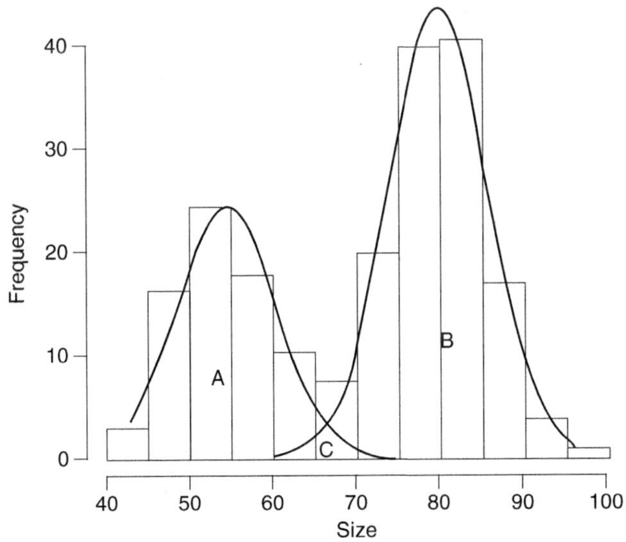

Fig. 10.1 Fictitious example of a two group larval data set. The two bell shaped curves correspond to the probability density functions of the model. The individuals in area A are classified as group I, for area B as group II while the individual sin area C are classified as belonging to the first (I) or the second group (II) according to their classification probability. On size alone a threshold probability must be defined above which the individual belongs to the respective group (modified from Benaglia et al. 2009)

$$\mathrm{Cr}_i = \frac{D_{(i-1)\to i} - D_{(i-2)\to(i-1)}}{D_{(i-2)\to(i-1)}}, \tag{2}$$

where Cr_i is the Crosby ratio at instar i; $D_{(i-1)\to i}$ is Dyar's ratio for the growth from instar $i-1$ to instar i. Crosby's ratio is a standardized form of Dyar's ratio and in this sense it provides a test for the detection of lacking instars (Goldson et al. 2001; Johnson and Williamson 2006).

The most serious obstacle in the determination of larval instars is that some characters show different Dyar's ratios throughout larval growth. Actually in several studies it was found that many morphological characters exhibit a variety of coefficients of determination in the linear regression of their maxima against reference characters such as the maximal cranial width (e.g., Johnson and Williamson 2006).

Methods for the Determination of Larval Instars

Mixture Models

Among the first uses of normal mixture models in entomology and paleobiology are the works of Goldson et al. (2001) and Hunt and Chapman (2001). In clustering work mixture models are meant as a set of many statistical distributions in the data each representing a different cluster or component. The number of clusters and the degree of match of the data to a particular set of components is determined by Bayes factors approximated by BIC (Fraley and Raftery 1998). The basic idea is that a growth pattern like this shown in Fig. 10.1 can be approximated through a series of normal distributions the number of which is the number of instars. In particular, according to Hunt and Chapman (2001) each sample can be represented by the probability density function of i normal distributions, namely.

$$f_{\mathrm{mix}} = \sum_{i=1}^{g} \pi_i f(\mu_i, \sigma_i^2), \tag{3}$$

where π_i is the mixing proportion of the original distribution to the i-th component of the distributions represented by (3); $f(\mu_i, \sigma_i^2)$ is the normal pdf with mean μ_i and variance σ_i^2. The original sample is split and the normal mixture model is fitted to the various subsets. A likelihood test is use to find if the differences between the subsets are significant (Goldson et al. 2001). Hunt and Chapman (2001) provided a procedure to test the significance of the split into more groups. In general this procedure involves the comparison of the two hypotheses – i.e., fewer vs. more groups – by parametric bootstrapping. An important aspect of this work is the suggestion that the rejection of a hypothesis with more groups may be due to the small statistical power usually present in samples with many groups. The authors suggested a three step bootstrap procedure to solve the problem by estimating the statistical power and they provide a computer program with a user guide (Hunt 2001). Actually the first problem met in studies involving many groups is very acute in paleontological studies where there is always a shortage of fossil specimens. In the following sections we shall analyze the *Ampyxina bellatula* set with other methods. All the presented clusters analyses support the predictions of Hunt and Chapman (2001) of seven instar groups.

Finite mixture models are important models that usually include clustering, regression and hypothesis testing functions. In analyzing the data sets used in this study three basic packages are employed derived from the libraries of the R language (R Development Core Team 2008). The package "mclust" (Fraley and Raftery 2007, 2009) is a set of computer routines suitable for normal mixture modeling and clustering based on specified models. It can provide also hierarchical clustering while it includes functions for the EM (Expectation-Maximization) estimation of mixture components and Bayesian Information criterion (BIC) for the evaluation of comprehensive cluster analysis. For the analysis of finite mixture models the package "mixtools" (Benaglia et al. 2009) includes many methods further to the more traditional ones like EM reflecting thus the progress made by recent research as mixture models are in the front line of statistical research. Among the packages that examine finite mixtures of regressions models "flexmix" (Gruen and Leisch 2008) was used for the Gaussian family of regression models while the package "mixreg" (Turner 2009) is a set of regression functions for easy fitting and visualization of the data. In these models the basic idea is that the problem of assignment to an instar is actually a problem of "multinomial logistic regression" since there is a predefined number of outcomes (=instar groups or classes) and each individual is assigned to a class. In this way the similarity of the regression models to the model-based clustering methods can be easily perceived.

Canonical Discriminant Analysis and Cluster Analysis

Classification by Canonical Discriminant Analysis (CDA) is used to achieve the most discriminative variables for the arrangement of samples in a space of reduced dimensionality in a way that maximizes the distances between the a priori formed groups and the independence between the axes of the configuration (Morrison 1984; Pimentel 1992; Johnson and Wichern 1998; Petrakis et al. 2008). The variables contributing most to the discrimination of preformed groups are shown by means of the *F-ratio* statistic as a criterion for inclusion or removal of the compound in a forward stepwise CDA mode. *Wilks'* λ, *Pillai's Trace* and their associated F approximations were used to check the significance and estimate the importance of each compound in CDA analysis (Engelman 1999). The performance of the method is measured by the percentage of total variance explained by all significant

Table 10.1 Classification matrix of CDA of *Ampyxina* data

a

Predicted ▷ Actual △	1	2	3	4	5	6	7	8	9	% Correct classifications
1	1	0	0	0	0	0	0	0	0	100
2	0	2	0	0	0	0	0	0	0	100
3	0	0	9	0	0	0	0	0	0	100
4	0	0	0	12	0	0	0	0	0	100
5	0	0	0	0	19	0	0	0	0	100
6	0	0	0	0	0	23	0	0	0	100
7	0	0	0	0	0	0	21	0	0	100
8	0	0	0	0	0	0	0	5	0	100
9	0	0	0	0	0	0	0	0	5	100
Total	1	2	9	12	19	23	21	5	5	100

b

Predicted ▷ Actual △	1	2	3	4	5	6	7	% Correct classifications
1	2	0	0	0	0	0	0	100
2	0	10	0	0	0	0	0	100
3	0	0	12	0	0	0	0	100
4	0	0	1	41	0	0	0	98
5	0	0	0	0	21	0	0	100
6	0	0	0	0	0	5	0	100
7	0	0	0	0	0	0	5	100
Total	2	10	13	41	21	5	5	99

The Jackknifed matrix is in general the same except for the outlier in panel (**a**) which is classified with the two individuals in group 2 lowering the total classification efficiency to 98%

discriminant axes. In some sense this is measure of non-linearity and complex interrelations among variables since CDA is a linear technique.

The classificatory efficiency of CDA can be seen in the resulting classification tables (e.g., Table 10.1a, b) which illustrate the group affiliation of the samples. The ability of samples to be predicted by the rest is shown in a jackknifed table (Table 10.1b) and is in general smaller than the one of the first. In the studies of instar groups of rare or extinct organisms, it is not uncommon to have very few individuals in one instar. The jackknifed table is then much smaller than the classification table which means that an entire instar is useless since the respective group cannot be defined in an adequate way.

For each sample the Mahalanobis multivariate distance is calculated from the group centroid in order to estimate the probability of misclassification in the predefined groups. Almost all statistical packages in the same table are predicting the new samples to be classified usually with a distinct mark. These new samples are not engaged in the calculation of classification functions of CDA. This feature of CDA reveals its ability to classify new individuals with known and measured size but unknown instar affiliation.

Many misuses of CDA can be seen in the related literature. Some of them are related to the violation of the assumption of multivariate normality which is rarely the case in instar size groups. Moreover the omission of an instar is very important since it violates all the underpinning ideas in the recovery of instar groups. This is faced by model-based clustering like *mclust* (Fraley and Raftery 2009) which considers data to be incomplete in the sense that some groups may be absent from the sample. As a result the discriminant space is configured in such a way as to accommodate only existing size groups. Researchers on CDA use a cross-validation procedure to cope with this problem (Engelman 1999; Fraley and Raftery 2007). According to this procedure, the original sample is used as the *learning set* while the unknown samples are considered as the *test set* in a way very similar to neural network based clustering.

The classification matrix of CDA predicts its ability to estimate the group affiliation of unknown samples such as the suspected for adulteration oil samples. In this way the method resembles the classification by the means of artificial neural networks but it is superior to it because of (1) the visibility of the discriminant space, (2) the detection of trends in the data since the samples are ordinate in a maximum variance way in the most parsimonious dimensionality, (3) the inclusion of the most significant size variables in the algorithm, (4) the statistical control of the dimensionality, and the group affiliation of unknown samples, and (5) the control of the researcher in the analysis.

Classification and Regression Trees

Since CDA is a linear method and requires the definition of original group affiliation with central objective to ordinate the samples (Oksanen et al. 2008) the Classification and Binary Trees (CBTs) of CART methodology we employed as another, generally not complementary method capable to handle non-linearity and any possible interaction of size dependence, in order to produce a set of simple rules, which identify existing samples and with this information find the group affiliation of any new sample (Breiman et al. 1984; Wilkinson 1985, 1999; Steinberg and Colla 1997). The method has been used for the classification of three species (74 individuals) of the flea-beetle genus *Chaetocnema* (Taylor and Silverman 1993). Morphological data for this species set was provided by Lubischew (1962) who also classified the flea-beetles with CDA. The importance of classification and regression trees in ecological work and its superiority to linear models was shown by De'ath and Fabricius (2000). More precisely, they showed that linear models failed to find patterns revealed by CBTs.

In order to produce a set of simple rules in the form of inequalities, which will identify the origin of any new sample, is essential in any larval instar detection methodology. In addition, while CDA predicts class affiliation by using a linear combination of all size variables engaged in the construction of classification equations CBTs do not require linearity. For instance, if larval groups have a weird non convex shape – e.g., "kidney" shaped – as a result of uncorrelated size measurements shaped arrangements in touch they cannot be separated by linear combinations of size morphometric variables (Johnson and Wichern 1998).

The CBT method possesses some features that make it superior to any other classification method (Breiman et al. 1984; De'ath and Fabricius 2000). Namely:

1. It is unaffected by outliers, which at the end (terminal leaves) are forming their own groups; this property is very important in studying larval instars since many outliers may occur in a sample due to sampling stochasticity. In particular paleontological studies possess such outliers (e.g., Hunt and Chapman 2001) since a particular instar, especially early instars or developmental stages of an instar with short duration are not equally available to fossilization as other instars and may not appear in a particular record.
2. The variables can be either continuous or categorical – i.e., some instar affiliations may be unambiguous as they are coming from rearing of larvae – and also some samples – i.e., larval remains as in Kishi (1971) see also Johnson and Williamson (2006) – with missing information on some size variables or some compounds – i.e., if chemical composition of cuticular extract is studied (Page et al. 1990, 2002) – can be processed with effect shown only in their probability of class affiliation; this is achieved by using alternative surrogates which are used in the case the original compound is missing.
3. It is inherent in the algorithm of the construction method that a stopping-rule is not necessary because of the "tree pruning" which is employed in order to produce the "best tree".
4. The method has no any specific strategy to split the data and as a result a compound can appear in splits at different nodes of the tree; this is a property that renders the method immune to the models underlying the size of larvae. This is welcome when we have no idea on the underlying model of the size variables but in the case of invertebrate instar groups the biological appealing Gaussian bell shaped growth curves predominates all studies.
5. Any type of context dependence or interaction of compounds and any non-linearity in the data can be handled properly. For instance, the cranial widths and the maximal width near the insertion of the cranium into the prothorax of the late larval instars of *Saperda vestita* (Coleoptera, Cerambycidae) are probably highly correlated (Johnson and Williamson 2006). CART takes into account this fact and uses only the most suitable size variable at a specific split. In a further split the algorithm may use this variable or another one among the highly correlated ones.
6. Simulation studies have shown that CBTs perform 10–15% better in comparison with logistic regressions and CDA (Steinberg and Colla 1997). However, there no simulations for the performance of CBTs in the case of just one size variable such as the first principal component of cranial length and cranial widths of the trilobite *A. bellatula* (Trilobita) (Hunt and Chapman 2001).
7. CART methodology provides a set of important size variables and a specific weight for each one even if they never appear in the tree since it keeps track of all surrogate splits in the tree growing procedure without regard to the subsequent

tree pruning that removes some splits. The immediate result is that some variables such as the type of leaf mines of the olive kernel moth *Prays oleae* (Petrakis 2000), the concentration pulses of the released ectydsteroid hormone in *Cotesia congrecata* (Hymenoptera Braconidae) (Gelman et al. 1999) or the width of the linear leaf mines of *Liriomyza sativae* (Diptera, Agromyzidae) (Jiao et al. 1998) may not appear at all in the set of splitting variables but they are influence through their weight.

8. The tree grows according to splits that produce maximally informative and "pure", according to an impurity function. The reduction of error in the classification is monitored by means of a loss function. Several loss functions have been proposed (Breiman et al. 1984) and in this study the splitting method has been made through the information theory compatible *Gini–Simpson* index in the form of "towing" a term coined by Breiman et al. (1984) on the basis of the better reduction in error achieved. Taylor and Silverman (1993) recognized some limitations of the *Gini–Simpson* splitting criterion. They found that it is biased towards equally sized subsets as a result of the concentration on the purity of offspring branches.

Worked Examples

Instars in Prays oleae

The larval instars in the olive kernel moth (*P. oleae*, Lepidoptera, Yponomeutidae) were studies for the leaf generation in a variety of olive tree cultivars and varieties (Petrakis 2000). In parallel to the measurements of the epicranial larval sclerites insect rearing were set for each leaf substrate as a reference (Fig. 10.2). From these measurements approximately 1,000 larval individuals were classified to predefined instar groups through by means of classical CDA.

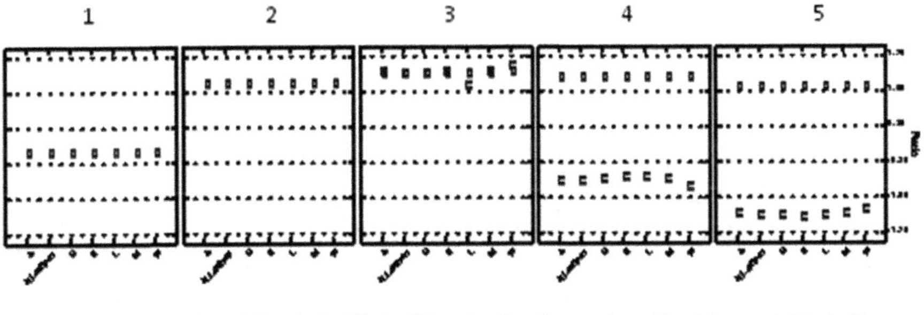

Fig. 10.2 Dyar's (D) and Crosby's (C) (×10) ratios for *Prays oleae* (Lepidoptera). Each diagram corresponds to a specific instar. The abscissa shows the *Olea europaea* cultivars and the wild variety (=var. *sylvestris*). "All cultivar" corresponds to the mean of all cultivars. Cultivar names are denoted by single letters (A: Amphissis, G: Megareitiki, M: Manaki, K: Koronaiki, L: Ladoelia, W: wild variety var. *sylvestris*)

The larval instars show the same Dyar's ratio in all groups and all substrates but Crosby's ratios while remain more or less consistent within the various substrates in all instars except the lower value on Ladoelia (L) in L3 they change sign in L4 and L5 indicating that the standardized growth rate is lower in advanced instars. A much smaller variation of the growth rate among instars was found in the trilobite *A. bellatula* by Hunt and Chapman (2001). The difference with this study is that the clusters found there correspond presumably to instars. The authors discuss the variation of the growth rate on the basis of existing instar groups. According to their suggestion should they split their group I into two groups the growth rate would have been much smoother (~1.12 and 1.10). However, this grouping violates the proposed method for the grouping of instars (many more details can be found in their paper). The important point is that in the case of *P. oleae* size groups correspond to larval instars. Surprisingly, in cases where this is not verified the growth rate – or equivalently the Dyar's and Crosby's ratios – is not incorporated in the clustering algorithm. Instead, growth rates are invoked as a posteriori verification of the grouping.

Employing discriminant analysis taking also the second size dimension of epicranial sclerites larval groups were recovered in a highly significant analysis ($\lambda_{\text{Wilks's}} = 0.047$, $F_{\text{approx}} = 610.37$, $df_1 = 8$, $df_2 = 1{,}350$, $P < 10^{-5}$ and $T_{\text{Pillai's}} = 0.95$, $F_{\text{approx}} = 153.863$, $df_1 = 8$, $df_2 = 1{,}352$, $P < 10^{-5}$). However, the classification of individual larvae is not perfect (91%) as it can be seen in Fig. 10.1. The misclassifications are due to the wild olive tree variety which is an inferior feeding substrate. On the other hand it can be seen that larvae derived from various cultivars form more or less tight clusters.

Submitting the same data set for *P. oleae* with only the measurements of the width of larval epicranial sclerites in a mixture models analysis – through the "mclust" program (Fraley and Raftery 2009) we found also a five-component set of the original set of larvae comprising a five-instar (=group) set of measurements. Interestingly, 8% of the larvae cannot be assigned to a specific group (=instar). This happened because size was not close to the mean of a certain group so that the a posteriori probability of group affiliation was not large enough for a safe assignment or a specific larva can be assigned to two groups (area C in Fig. 10.1; bottom panels in Fig. 10.4). At this point a distinction must be made between CDA (Fig. 10.3) and mixture models (Fig. 10.4). In CDA the groups are predefined while in mixture models the number of groups is sought and in general depends on the number of components – i.e., model functions such as Gaussian, binomial or poisson – that are mixed in the data (Fraley and Raftery 2009).

The analysis of this data set with the "mixtools" package specialized in mixture models regressions (Benaglia et al. 2009) involving the "`normalmixEM`" function which uses the EM algorithm to find a local maximum of the likelihood surface provided poor fit when the dependent variable was y = 4, 5 or 6 groups. The analysis was done in the context of multinomial logit regression. The calculation and the plot of the estimated statistical distributions gave a two or three component set of functions (see also Elmore et al. 2004). It might be that this package is not suitable for this type of biological work although it has been used in very complicated series

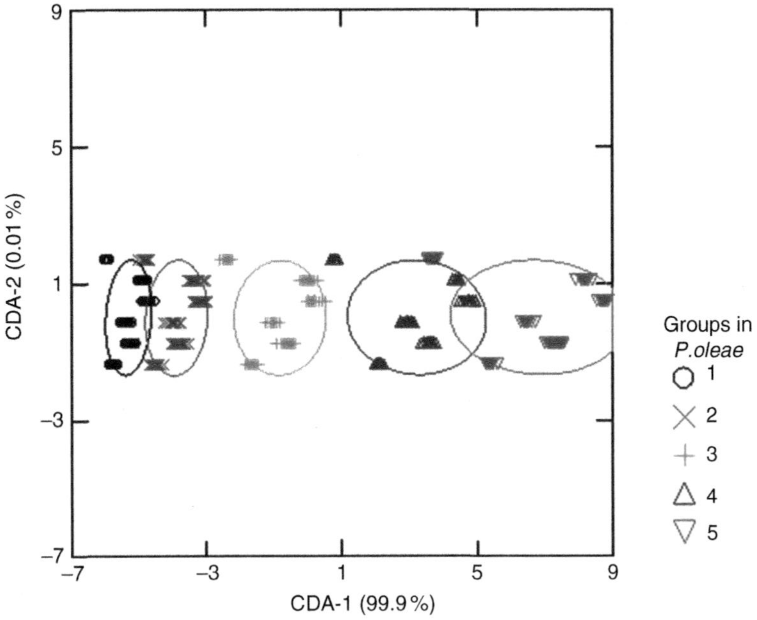

Fig. 10.3 CDA diagram of *P. oleae* data. The larval groups are five while the entire variation is accounted for by the first discriminant axis (99.9%)

of experiments. The general methodology in *mixtools* package considers the mixture problem as an incomplete data set where the actual data configure the "complete" sample space and another sample space that consists of the "incomplete" observations. In this respect it resembles the methodology of Hunt and Chapman (2001) where a sample is generated on the basis of parameter estimates of hypothesis of fewer groups (H1 according to the authors).

Instars in Ampyxina bellatula

This data set is based on the measurements of cephalic length and width taken by Brezinski (1986) from 97 individuals of the Upper Ordovician trilobite *A. bellatula* (Trilobita, Phiophorida).

This data set was analyzed by Hunt and Chapman (2001) in a novel approach in finding the number of instar groups in the data. Their method involves mixture models and maximum likelihood estimators for the finding of distribution parameters on the basis that the measured variables are distributed normally within each instar. The computation is automated by a computer program written by the first author (Hunt 2001) and distributed by him upon request. The diagram of the two cephalic dimensions (cephalon length and cepahalon width, Fig. 10.18) is shown in

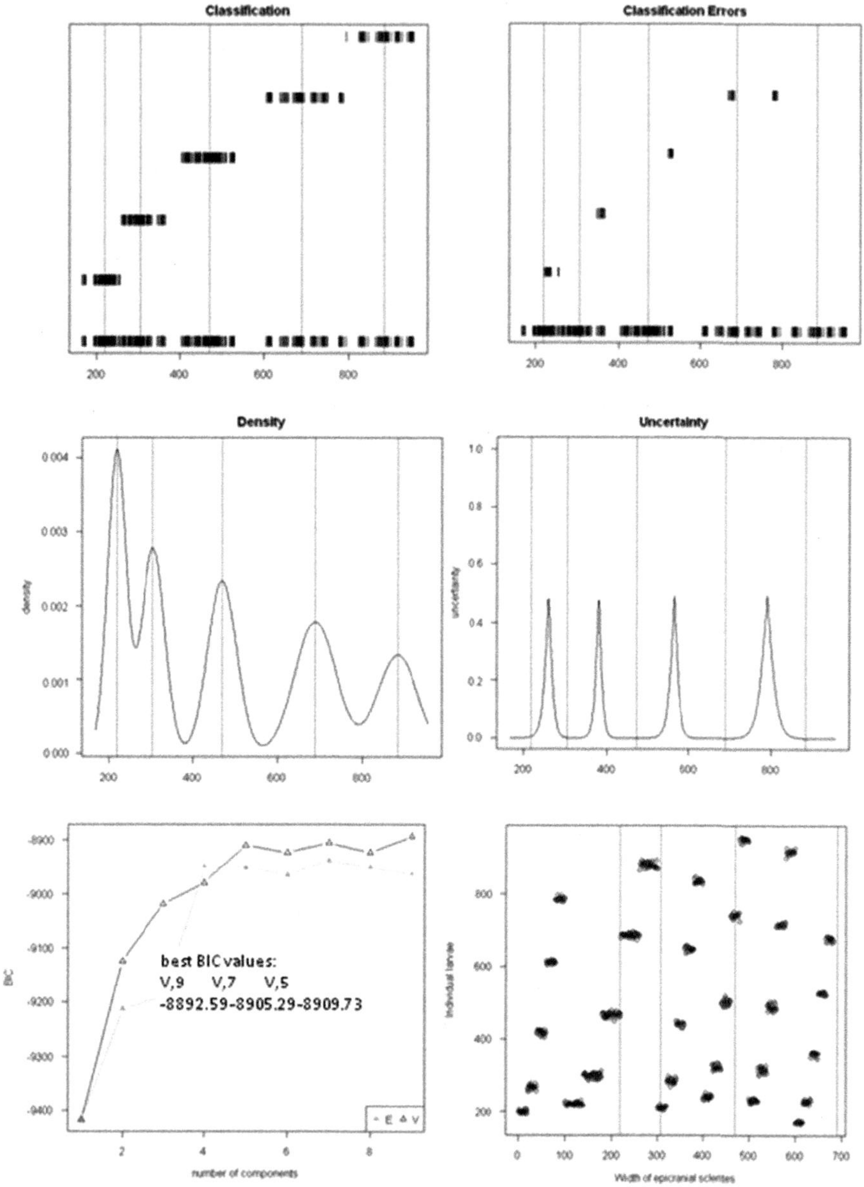

Fig. 10.4 Model-based clustering of *P. oleae* data. Each individual larva is represented by a small vertical bar. *Upper left*: The specimens at the bottom of the plot are classified in the five groups at a height that corresponds to the mean dimension of each instar. *Upper right*: The classification errors of the classification at the right. It must be noticed that the fifth group lies at the height of the fourth group. *Middle left*: The proportion of larvae in each instar (=group) is shown on the y-axis. The smoothing is derived from the algorithm of the package "mclust" on the basis of the number of clusters judged by the BIC criterion. The width of the bell shaped curves varies because the model was of the varying variance type. *Middle right*: The uncertainties of the assignment of each larva to

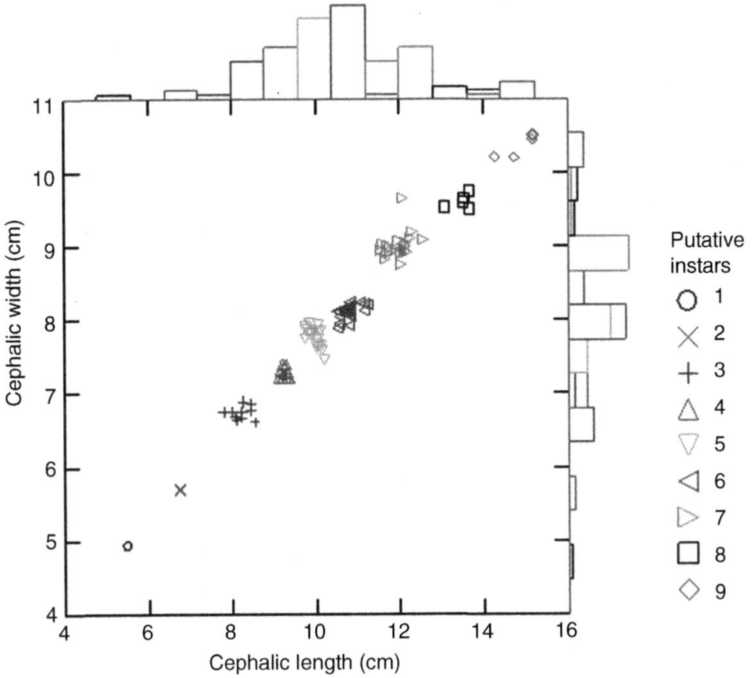

Fig. 10.5 Scattergram of the cephalic length vs. cephalic width of the *Ampyxina* (98 recognized individuals). The diagram is bordered by the histograms of each group out of the nine recognized by inspection of the clusters. The nine histograms can be seen that intermingle and have different mean and standard deviation. Moreover, the two dimensions have generally different histogram parameters

Fig. 10.5 where nine groups are easily seen. Brezinski (1986) on the basis of putative clusters interpreted them as instars and visualized the data as a set of eight clusters leaving out one small specimen that forms its own cluster. Hunt and Chapman considered this and in addition they united all large specimens in one instar-group forming seven clusters. They found that the number of seven groups is favored by their maximum likelihood approach. The important point in this method is the statistical testing of the number of groups.

Fig. 10.4 (continued) the revealed groups on the basis of the a posteriori probabilities estimated by "mclust". *Bottom left*: BIC values for all components of the *mclust* classification of the *P. oleae* instar data. Since there is only one measured size variable – i.e., epicranial sclerite width – only two models are possible (E: equal volume of the components, V: variable volume of the components). Among them are the V model 9, 7, and 5 components. Evidently the 5 components model is preferred since it is known that there are five instars in this insect but the 7 and 9 component grouping can be also considered. *Bottom right*: The densities resulting from this clustering produced the scattergram the diagram with many more clusters but the vertical lines indicate that some clusters cannot be separated on the basis of mixture modeling

The approach through mixture models is fully justified as shows the border histogram of Fig. 10.5 of the individuals shown in the diagram. Many distributions are seen intermingled for each putative instar group and is expected a mixture model based clustering will unfold the bell shaped Gaussian curves that pertain to each group. Indeed, submitting the same data set in a mixture model based clustering resulted in BCI and likelihood favored set of seven components (=groups) although the Akaike Information Criterion (AIC) and the corrected from of it (CAIC) favored the two component solution. It must be noted that AIC and CAIC are more sensitive – i.e., more conservative – to the number of parameters and for this they prefer the simpler model which is not wanted in this type of problems. On the contrary BIC as Bayes' factor estimator (Fraley and Raftery 1998) is more sensitive to sum-of-squared-deviations from the observed data (=less conservative). On the other hand the hypothesis of seven groups holds true for a variety of biological reasons (Hunt and Chapman 2001) that is the seven groups solution makes the growth ratios between successive groups consistent with the values found in other extant arthropod assemblages and in addition they vary in the narrow range (1.12–1.08; Dyar's ratios fluctuated around the mean value 1.15 according to the conventional clustering with Ward joining of Euclidean distances).

Employing the *mclust* method, the BIC value (33.34) of the fitted model of seven clusters is the best among all grouping numbers in the interval [6, ..., 10]. More importantly the fit to all other models was unsuccessful for one or both measured variables (cephalon length, cephalon width). The next best model predicts eight clusters with BIC value 47.39. As a rule of thumb given by Fraley and Raftery (1998, 2009), this indicates strong evidence of difference between the two groupings, presented as an alternative method of statistical testing the number of groups. On the other hand the fit of the seven cluster model was highly significant at all variable coefficients except for "cephalon width" in cluster 6 and "cephalon length" in cluster 2. However both coefficients were very small – i.e., less than 0.00001% of the minimum value in other clusters. The number of seven clusters is in agreement with the instar grouping level found by Hunt and Chapman (2001) with a difference that they used the principal component of the two measurements while here both variables are engaged, an additional information loss. It must be noted that the program written by one of them handles only one size variable. In addition *mclust* methodology statistically tests the importance of the variables in finding whether the measured variables within each instar are normally distributed, a feature with appealing philosophical background for invertebrate biologists. The methodological question of having fewer individuals in large instar numbers, due to survivorship processes, and in effect lower power in the employed statistical tests underlies all paleontological studies and sometimes the work on extant insects.

Performing CDA on the morphometrics data of *A. bellatula* the problem of the number of instars enters at the first phase because discriminant analysis functions on predefined groups. Hence the classification ability of CDA can be visualized as a verification process of the a priori selection of the instar affiliation of individual measurements (Johnson and Wichern 1998). However, it is risky to rely on the significance or the classification ability of CDA since significance can be achieved by a varying number of initial groups. In Fig. 10.6 is shown the diagram of the

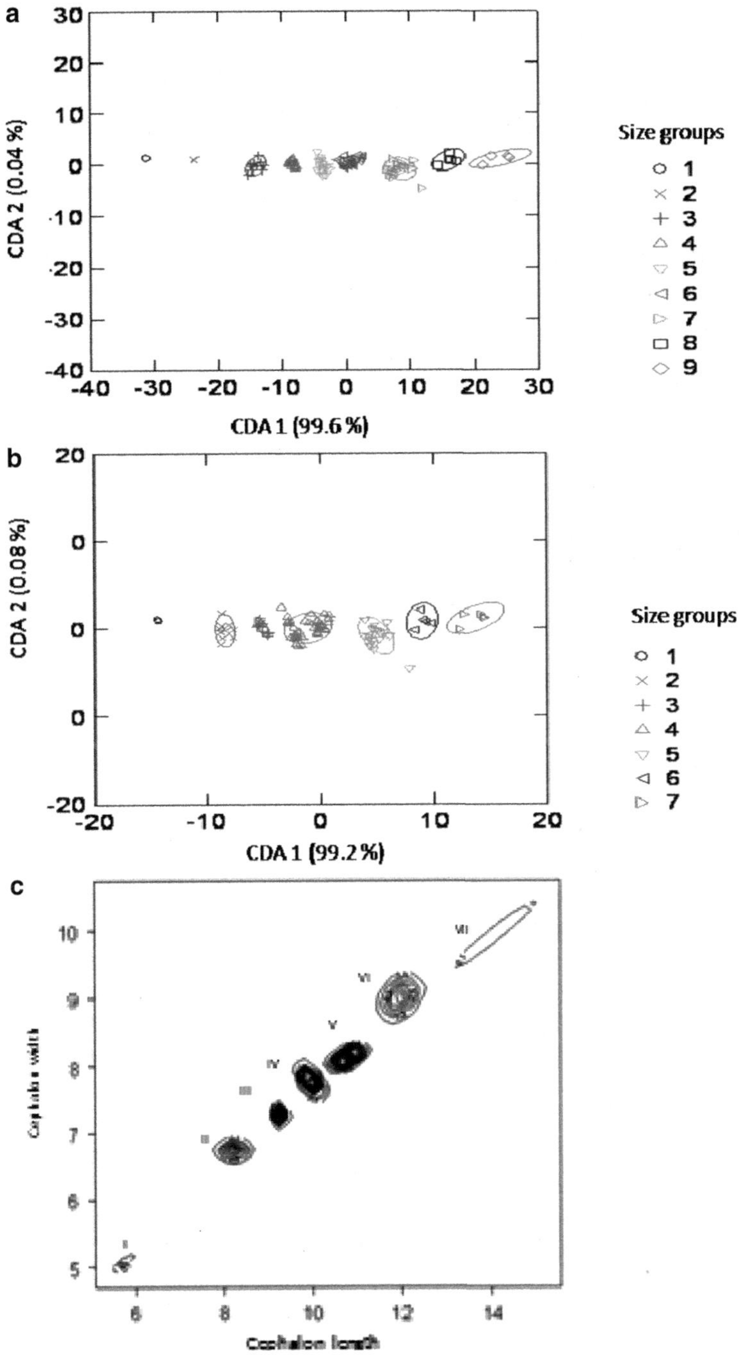

Fig. 10.6 Diagram of the CDA of *Ampyxina* data. (**a**) Nine groups circumscribed by 95% ellipses. (**b**) Seven groups and 95% ellipses. The misclassified individual in the third group from the right can be easily seen. (**c**) Density contour plot of the seven clusters coming from the Gaussian model clustering of *mclust* R library. It is obvious the lack of an instar between I and II

discriminant scores of two highly significant analyses ($P < 10^{-5}$) of *A. bellatula* data. The classificatory efficiency of the analysis (on nine [**a**] and seven [**b**] groups) is shown in Table 10.1. The second panel **b** shows the same classification as panel **a** with one species left outside the analysis – i.e., the outlier in panel **a** which was removed in panel **b** – and the groups 5 and 6 which were fused in group 4. In panel **a** the classification is perfect but in panel **b** one individual in group 4 was predicted to be member of group 3. Thus the number of groups predicted from CDA is different from what mixture model-based clustering predicts. The reason for this does not lie in any violation of the assumptions of CDA such as the normality, continuity, and commensurability of descriptors. It is rather a problem of the low number of individuals that cover at different degrees the gaps between the instars. In other words, the overlap of groups varies among instars.

Since the discriminant axes are only a summary of the morphometric sample, the first few discriminant axes carry the most important information content of the original data. In this data set only two original measurements were taken and in effect the entire information is contained within the discriminant score configuration of individuals. On these premises it is feasible to select the number of groups on the basis of classification. Evidently, the larger the number of groups the better becomes the classification efficiency since the individuals are described more efficiently. The number of groups is hence the largest that results in suboptimal classification (optimal classification is 100%).

The *A. bellatula* data set has only two measured variables – i.e., cranidium length and cranidium width – taken on the cranidium part of the body and these measurements are highly correlated. A hierarchical cluster analysis (HCA) of this set can reveal size groups depending on the level at which the groups are considered. An example is given in Fig. 10.7 involving Ward joining of Euclidean distances. The set was analyzed in SYSTAT (Wilkinson 1999) and PAST (Hammer et al. 2009) gave exactly the same results.

The group numbers in CA is validated with a set of functions shown in Fig. 10.7b for various clustering levels. The number of seven clusters is supported by all three plots. At this level σ_{rms} exhibits a local peak which is also shown by pseudo-F function. In the third diagram pseudo-t is strongly peaked at this level. In all diagrams the level of seven clusters is supported. However, it must be kept in mind that these cluster validation statistics are not conventional *F*- or *t*-tests since the assumptions for these statistics do not hold true in clusters since the data coming are not random samples.

It has been reported that Dyar's ratio is valid for many organisms [for fossil arthropods (Kopaska-Merkel 1981); for trilobites (Hunt and Chapman 2001); many insect species (Daly 1985)]. This is also in agreement with the life history traits and the ecology of trilobites (Cisne 1973; Chatterton et al. 1994). For *A. bellatula* the diagram shown in Fig. 10.8 shows the Dyar's and Crosby's ratios for a number of instar size groups (5–9). It can be seen that the largest number of groups that the Dyar's ratio stabilizes at value 1.15 is 7. Above it there is an excessive number of groups while Crosby's ratios varies a little between −0.12 and 0.02. Crosby's ratios are prevalently negative which indicates that advanced instars have a lower growth

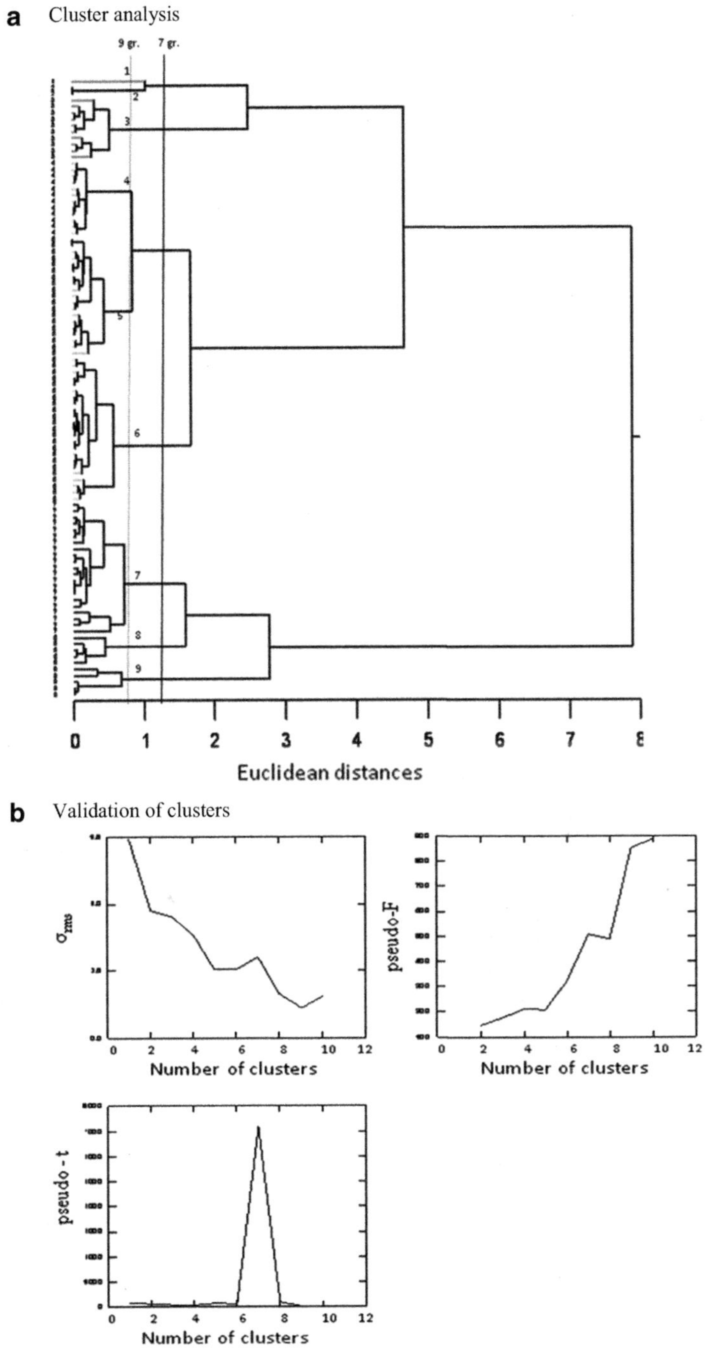

Fig. 10.7 (continued)

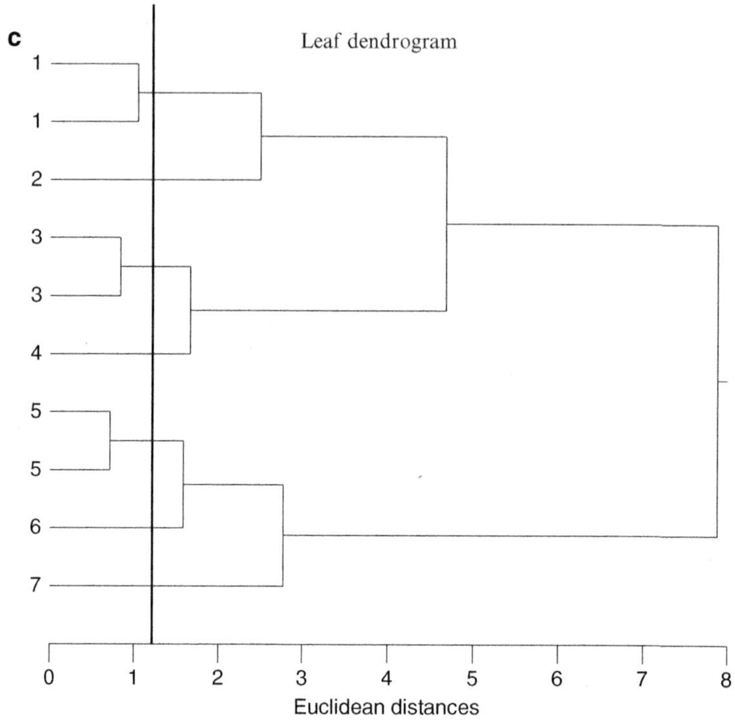

Fig. 10.7 Dendrogram of cephalic dimensions of the trilobite *Ampyxina* as measured by Brezinski (1986) and used in the detection of the number of instars by Hunt and Chapman (2001). In plot (**a**) the dendrogram of Euclidean distances in a "Ward minimum variance" joining algorithm is presented. The *numbers at the left* shows the putative instar as detected by inspection of the data. The *two lines* correspond to the level of seven clusters (the *dotted line* is drawn after the first sample is removed). In plot (**b**) are presented the validation indices σ_{rms}, pseudo-F and pseudo-t are depicted to give a rough idea about the statistically significant clustering level (here is the level of seven clusters). The leaf dendrogram of plot (**a**) is shown in plot (**c**) at the level of ten leaves. The *vertical continuous line* corresponds to the level of seven clusters. Note that the group at the leaf 1 consists from the outlier sample 1

rate. This has been exploited by Hunt and Chapman (2001) and is supported by extant insects (Johnson and Williamson 2006).

The *CART* methodology produces the classification shown in Fig. 10.9. The dendrogram includes the smallest individual out of 98 individuals in a separate cluster. If this individual is removed then an eight cluster configuration is produced, the 6.75 splitting value is removed and all remaining values are unchanged. The employed index is the Gini–Simpson one while the proportional reduction in error is 1.00. All terminal nodes are genuine in the sense that they contain no impurities and individuals from only one cluster. The SYSTAT software was used for the presentation while validation, complexity and examinations of surrogate splits were done through the *rpart* package (Therneau and Atkinson 2008) in R library (R Development Core Team 2008).

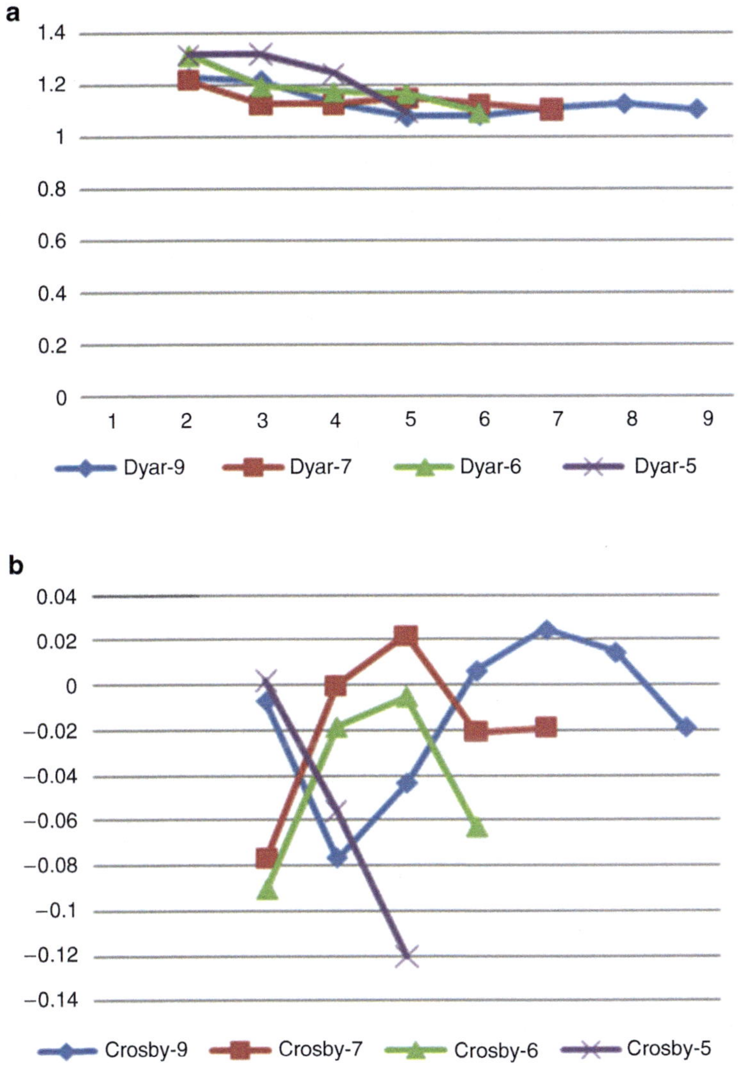

Fig. 10.8 (**a**) Dyar's ratios and (**b**) Crosby's ratios for *Ampyxina bellatula* (Trilobita) instars under various instar numbers (showed in the legend beneath diagram a. The numbers of the groups is the abscissa of both diagrams. The differences in the *y*-scales must be noted

All numbers of groups up to in the range [5, . . ., 10] gave the same partitioning as the one presented in Fig. 10.9 with the left branch of the tree accordingly pruned. The corresponding instar groups are fused according to the new grouping. This point can be clarified by inspecting the plot in Fig. 10.20. The number of seven instar groups is selected because it presents the smallest departures from an increasing trend of the standard deviation (σ) or the standardized *CV* (Fig. 10.10) (for related literature see Hunt and Chapman 2001).

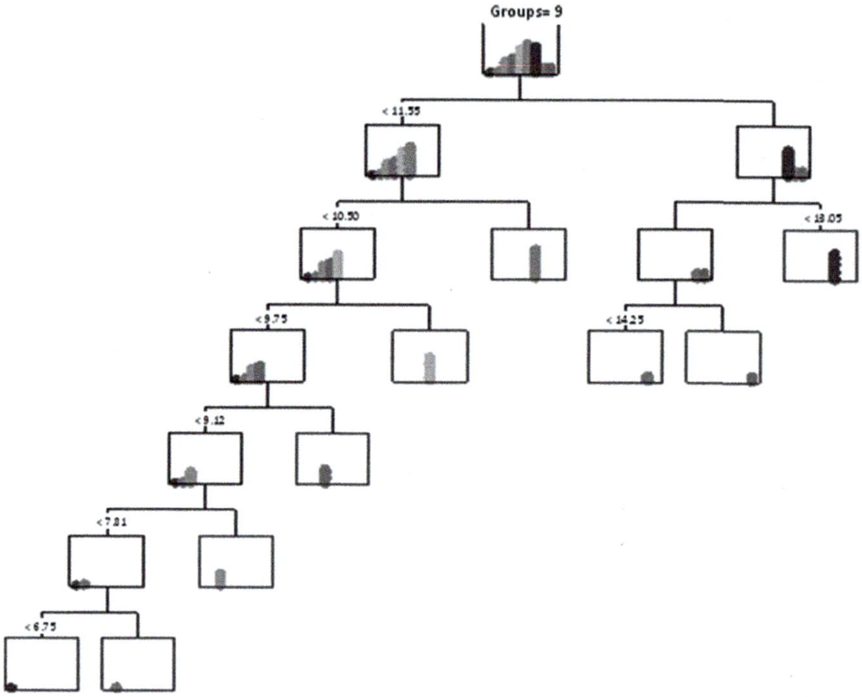

Fig. 10.9 Classification binary tree (mobile) of *A. bellatula* cephalic length and width at the level of nine recognized dimensional groups (presumably instars). The number of groups is nine and includes the outlier individual in a separate cluster. Only the variable "cephalic length" determined the hierarchical structure

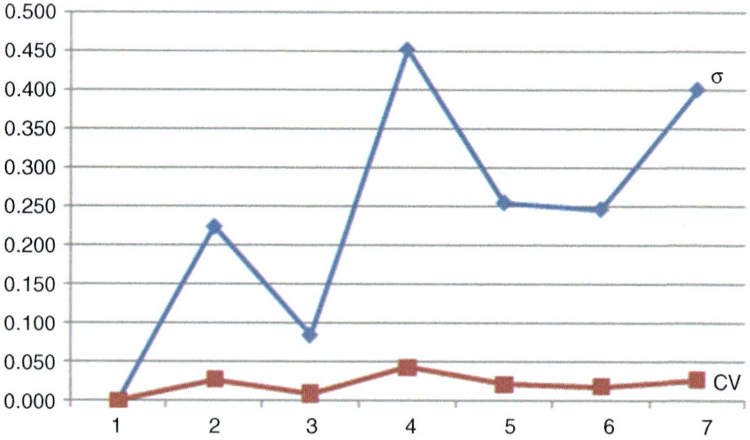

Fig. 10.10 Classification binary tree (mobile) of *A. bellatula* cephalic length and width at the level of nine recognized dimensional groups (presumably instars). The number of groups is nine and includes the outlier individual in a separate cluster. Only the variable "cephalic length" determined the hierarchical structure

Instars in Piochaspis sellata

The trilobite *P. sellata* (Trilobita, Ptychopariida) comes from the collections of the Field Museum of Natural History (see above) and the number of instar groups in a sample of 38 individuals was analyzed in Hunt and Chapman (2001). Two specimens, namely the smallest and the largest individuals were excluded as outliers. On the basis of the maximum likelihood methodology, they found in the remaining 36 individuals and the bootstrap criterion for testing the fewer vs. many clusters that the four group model is favored over the one with three groups. However, too few individuals exist within each group in the case of four groups. This weakened the statistical power of the test and the authors preferred to consider three groups of instars within the data.

The same set of 38 individuals was submitted to a CBT analysis using functions in the *rpart* R library (Fig. 10.11). It was found that the four group model is produced

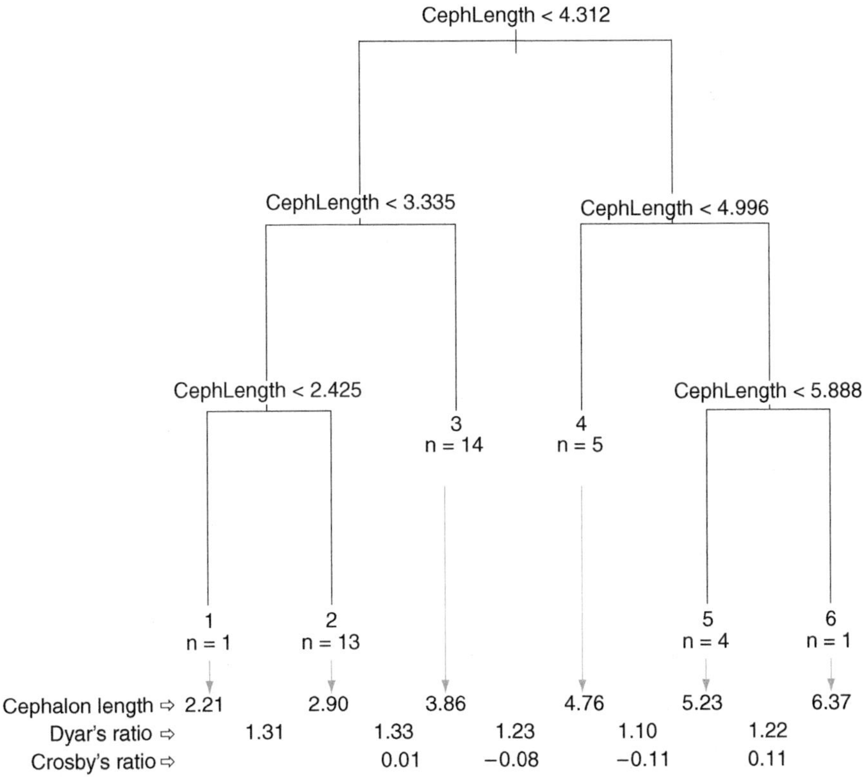

Fig. 10.11 Classification binary tree (mobile) on the basis of *P. sellata* cephalon length at the level of six groups (presumably instars). Below each terminal leaf (=instar group) is presented the mean value of cephalon length while in the transitional time space between instars are given the Dyar's and Crosby's ratios. The *dashed arrows* indicate the mean value correspondence and the values at each node indicate the cut values of the splitting variable which in all nodes is cephalon length. The number of individuals in each group is given by *n*

if the tree construction is parameterized with a minimum number of individuals in each terminal leaf being equal to one. After this there is no further split. A complexity parameter is predefined and indicates the lowest reduction of error below of which the split is likely to be pruned off by cross-validation. The number of groups is six because two groups include the smallest and the largest cephalon length. In general the other clusters coincide with those found by Hunt and Chapman (2001) except for the instar III of them which is divided in two by the cut value 4.996. It seems that CBT method forms the clusters on recognized peaks of the frequency distribution. In this set the small "gap" in group III is sufficient for the split into two groups (4 and 5 in Fig. 10.11).

Instar groups employing *mclust* philosophy (Fraley and Raftery 2007, 2009) gave exactly the number of instar groups of Hunt and Chapman (2001) in a reduced set of data when the smallest and the largest trilobite were removed. However the results are statistically confirmed by the means of the BIC value. These values ($E = -87.03$ and $V = -92.02$) correspond to the best models for the equal (E) and variable (V) volume model (Fig. 10.12).

Instars in Saperda vestita

The number of instars and the way that ten morphological characters can be used as indicators of larval stages in the linden borer *S. vestita* (Coleoptera, Cerambycidae) were studied by Johnson and Williamson (2006) in southern Wisconsin. A set of 1,532 larvae was removed from trunks of linden trees and measured under a compound microscope equipped with an eyepiece micrometer (Fig. 10.19 and Table 10.1). All measurements were taken to an accuracy of 10 μm. Due to removal procedures some characters were not observable. After the exclusion of these larvae 1,497 larvae remained and were analyzed (data provided by R. Chris Williamson). The authors also give a table of variable ranges useful for instar assignment of new larvae (Tables 10.2 and 10.3).

This insect group has five or six instars and the authors found that some variables are more reliable than others in determining the number of instars. They concluded that *S. vestita* has five instars but the possibility of six instars cannot be excluded due to the variability of measurements which may be a result of overlapping generations and sexual dimorphism.

On the basis of the different variables measured, the number of larvae in each instar is different and Table 10.3 is constructed using the predictions of the authors about the ranges of variables A, F and B in the instars. The number of larvae can be seen to vary extensively although the instar bounds were carefully selected and slightly changed by the authors (Johnson and Williamson 2006) to account for the overlap of instar groups. On the basis of frequency distributions (histograms of measured variables) and their reliability in conforming the Dyar's and Crosby's ratios they found that variables F, G, E and I were the most important predictors of instar-group affiliation.

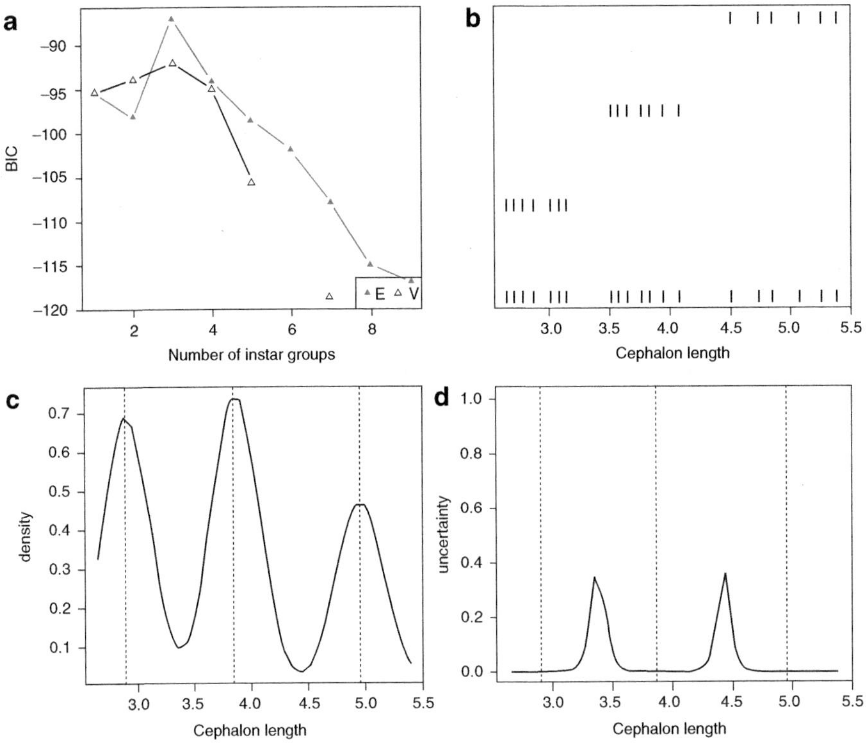

Fig. 10.12 (**a**) BIC values for various numbers of groups in the *P. sellata* data set by employing routines from *mclust* library of R language. (**b**) The classification of individuals in three groups since this is the favored number of instar groups. (**c**) Density plot and Gaussian estimated curves in the three clusters shown by the *dashed vertical lines* passing from the peaks of the *bell shaped curves*. (**d**) Uncertainty plot of individuals between the Gaussian curves. The probability of assignment of each individual is close to unity for those located near the peak and smaller as we proceed further from the peak

Table 10.2 Elements of larval morphology measured and their symbol designation

Symbol	Description
A	Left labial palpomere length
B	First thoracic spiracle length
C	First thoracic spiracle width
BC	First thoracic spiracle length/width
D	Base of labrum
E	Base of clypeus
F	Distance of antennal insertions
G	Head capsule width
H	Maximal prothoracic width
I	Maximal prothoracic width
J	Distance of labrum to prothorax

Table 10.3 Number of individuals in each larval instar according to the given character ranges

Instar	A	F	B
1	189	20	230
2	478	72	299
3	427	440	330
4	390	708	407
5	13	257	176
6	0	0	55

Johnson and Williamson (2006)

The problem presented by individual variables is not simple although all variables incorporated either an increase in size (A–J) or shape (BC). The size and shape variables are not in the sense of Mosimann and Malley (1979) who defined them as those variables that obey the equations:

$$G(a\mathbf{v}) = a \cdot G(\mathbf{v}) \tag{4}$$

and

$$\mathbf{Z}(\mathbf{v}) = \frac{v}{G(\mathbf{v})} \tag{5}$$

for size (4) and shape (5), where $\alpha > 0$ is an arbitrary number and $\mathbf{v}$ is a k-dimensional vector with all coordinates positive. These equations describe the shape of the $\mathbf{v}$ eigenvector. In this sense the quantities $\sum v_i$, $\sum v_i^{2\frac{1}{2}}$, $(\prod v_i)^{\frac{1}{k}}$ are all size variables.

We used the CART algorithm for the construction of the classification binary tree with the Gini index as a loss function (Breiman et al. 1984; Therneau and Atkinson 2008, 2009). In the analysis we firstly used the entire variable set and in a second attempt we then excluded variable B from the independent character set. However, in both attempts the instar-grouping suggested by B was the dependent variable.

Both analyses were perfect in terms of the proportional reduction in error – an analogue of the coefficient of determination in classical regression methods. Most importantly the variable responsible for the split according to Johnson and Williamson (2006) emerged in the CBT mobile as the only variable responsible for the split at each node. This is in agreement with the authors' predictions though the splitting values are slightly different because the instar bounds were slightly changed to account for the unique categorization of larvae at an instar. The splitting variable B is found to have values 0.250, 0.349, 0.425, 0.534, 0.632 [they can be read in the left CBT (bottom top and left to right) in Fig. 10.13 at a 4× magnification] while they define as instar bounds the values 0.24, 0.34, 0.42, 0.53 and 0.63.

The terminal nodes (leaves) have zero impurity which means that all larvae are perfectly classified by the spiracle length (B). Thus 55 individuals (Table 10.3) were categorized in instar VI unlike the other characters which are used as classifiers for five instar grouping dendrograms. Certainly not all measured characters

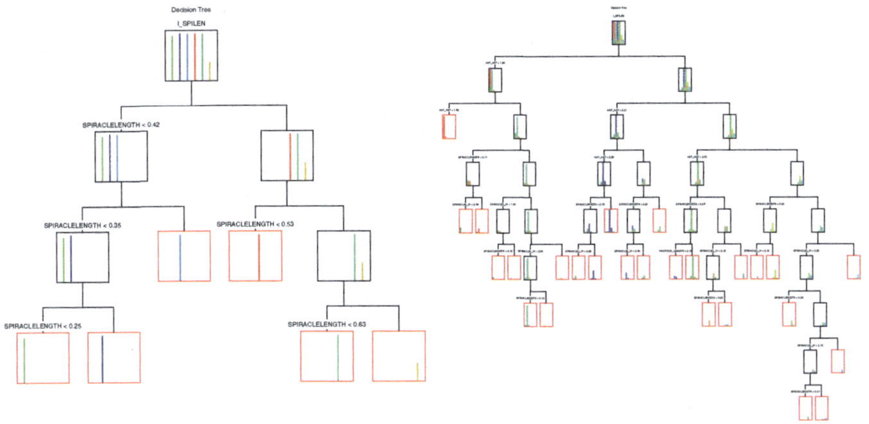

Fig. 10.13 Classification binary trees constructed with (*left*) and without (*right*) the inclusion of variable B (first thoracic spiracle length) in the set of eleven variables

produce the same grouping. The bounds of instar groups given in Johnson and Williamson's paper (2006) were specified in a way to conform to the Dyar's ratio. It is evident that the mathematically pure groups found here cannot account for the variability of the characters and the Dyar's ratio as was found by Johnson and Williamson who studied the larval growth of *S. vestita*.

Importantly, when the variable B is excluded from the independent variable set, the CBT (Fig. 10.13, right plot) produced diverges greatly from the variable set CBT with B left plot). The variables used in splitting the various nodes are F, C and BC in the order they appear in the mobile (Wilkinson 1985). The terminal leaves are not as a rule pure (impurity $= 0$) and they are much more numerous than the instars. This implies that variable A cannot be substituted by the others or a combination of them with the same effect on instar grouping.

The same picture emerges in all measured variables individually studied. The spiracle length is included because it shows another aspect of larval growth. The increase in size is shown by the spiracle width C while the change of the spiracle shape is accounted by variable BC which is the eccentricity of the spiracle's elliptic external outline. The shape is not a linear size dimension and in this respect it does not follow the Dyar's ratio. It varies inconsistently among instars. It takes the values $L1 = 1.99, L2 = 2.18, L3 = 2.25, L4 = 2.27, L5 = 2.33, L6 = 2.32$ according to instar affiliation as judged by the spiracle length. This indicates that the eccentricity approaches to unity (circle) as larvae molt to the next instar.

To separate the effect of size and shape the usual approach is principal component analysis-PCA (Mosimann and Malley 1979; Scott 1979; Pimentel 1992; Jolicoeur 1999). The same data set was analyzed with PCA (Wilkinson 1999). As expected, the principal axis bears the variability related to size while the second axis explains the shape variability (variable BC in Table 10.4). The two eigenvalues [$\lambda_1 = 9.34$, $\lambda_2 = 0.98$ in Fig. 10.14 (scree plot)] are significantly different from zero and account for 93.85% of the total variation. Impressively,

Table 10.4 Component loadings of the measured variables for the first two principal components

		PC1	PC2
A	Left labial palpomere length	0.896	−0.010
B	First thoracic spiracle length	0.974	−0.057
C	First thoracic spiracle width	0.942	0.261
BC	First thoracic spiracle length/width	0.301	−0.953
D	Base of labrum	0.966	0.009
E	Base of clypeus	0.978	0.004
F	Distance of antennal insertions	0.990	0.014
G	Head width	0.986	0.027
H	Maximal prothoracic width	0.974	0.034
I	Maximal prothoracic width	0.972	0.021
J	Distance of labrum to prothorax	0.936	−0.002
Percentage variance explained		84.93	8.920

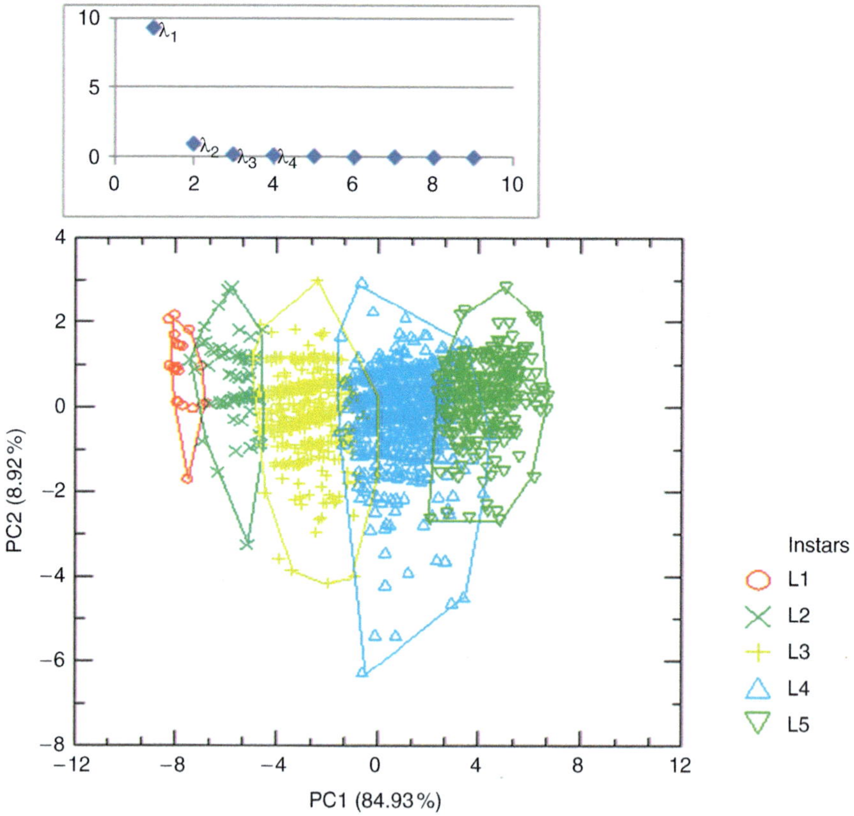

Fig. 10.14 The scree plot (*Upper*) and the principal component scores of all eleven larval variables. The five instar-clusters are indicated with the convex hulls around each cluster (*Lower*). The bulk of the variance in the original data is described by the first principal component (84.93%). The second principal axis (8.92% variance) is mainly describes the effect of the BC (thoracic spiracle length/width) shape variable

the first component (PC1) accounts for 84.93% and the second component (PC2) for 8.92%. It is commonly accepted that components $<10\%$ are not regarded important but here the importance is inferred from the scree plot. PC1 is highly correlated with all variables (except BC) while PC2 is highly correlated only with the variable BC (shape variable) (Table 10.4). The results of PCA are shown in Fig. 10.14 while the instars were determined on the basis of the best variable F (highest correlation with PC1 0.990). This agrees with Johnson and Williamson (2006) who found that F is among the most reliable morphological variables in the determination of instar groups. The inconsistency among variables emerges as an overlap between any two successive instars. The partial overlap between L2, L3, L4 and L5 complicates the assignment of larvae to instar groups.

To investigate further the problem of instar grouping model based clustering was applied. The rationale for this approach is outlined below.

1. Frequency distributions in the histograms of Johnson and Williamson (2006) show the existence of many normal distributions for all size variables.
2. Overlap between instar groups is shown in the principal components of size and shape variables (Fig. 10.14).
3. Biological explanations are given by Johnson and Williamson. Namely,
   (a) Sexual dimorphism in *S. vestita*
   (b) Overlapping generations
   (c) The life cycle which lasts 2 or 3 years
   (d) The overlap of measured variables between successive instars
   (e) The instars of other cerambycid insects
4. Larval instar grouping philosophy and the approach of Hunt and Chapman (2001) for extinct invertebrates (Trilobita).
5. Variation in size is caused by different feeding substrates of instars even in the same host plant individual.
6. Specialized methods and software availability (R Development Core Team 2008).

We used the model based clustering of the library mclust (Fraley and Raftery 2007, 2009) to analyze, test and depict the clustering results employed. Firstly, we used all measured morphometric variables in the independent set while the prior instar grouping was the one derived from the F variable (Fig. 10.15). The depiction of clusters was done for the variables F and E which are among the most reliable characters according to Johnson and Williamson (2006). The BIC value indicated seven instar groups as the best solution and the next best solution produced five instar groups. Seven groups is a biologically decoupled value therefore we adopted the five instar groups. The clusters are shown with 95% ellipses around cluster centroids. In Fig. 10.15 (upper plot) the entire point swarm is shown with 1,497 points.

However, there are many uncertainties and classification errors except at the ends of the diagonal, which shows that there is no agreement among the measured variables. In general, the entire picture is multidimensional and the depiction in a bivariate plot (F–E) does not succeed in revealing the instar grouping structure in *S. vestita*.

Fig. 10.15 The five components model produced by "mclust" according to all measured variables listed in Table 10.2. Only 2 out of 11 variables are shown in the diagram (base of clypeus and distance of antennal insertions) and for this there is extensive overlap of the presumed instar-clusters. The 95% confidence ellipses are drawn around the center of each cluster. *Upper*: Classification of all larvae on the basis of all morphometric data and five putative instar-clusters. Different shapes and grey-tones of symbols denote different clusters. The symbols (*open squares*) are shown to be widespread in the upper 2/3 of the diagonal line but it is separate from the other clusters in the hypervolume space constructed by all 11 variables. *Mid*: The same points are shown but now the symbols denote the probability of misclassification on the basis of size and darkness (the darker the point the closer to unity). *Lower*: Classification errors of various larvae (in the front) and clusters are shown as a background. The small and large dimensions contain the least misclassifications and errors

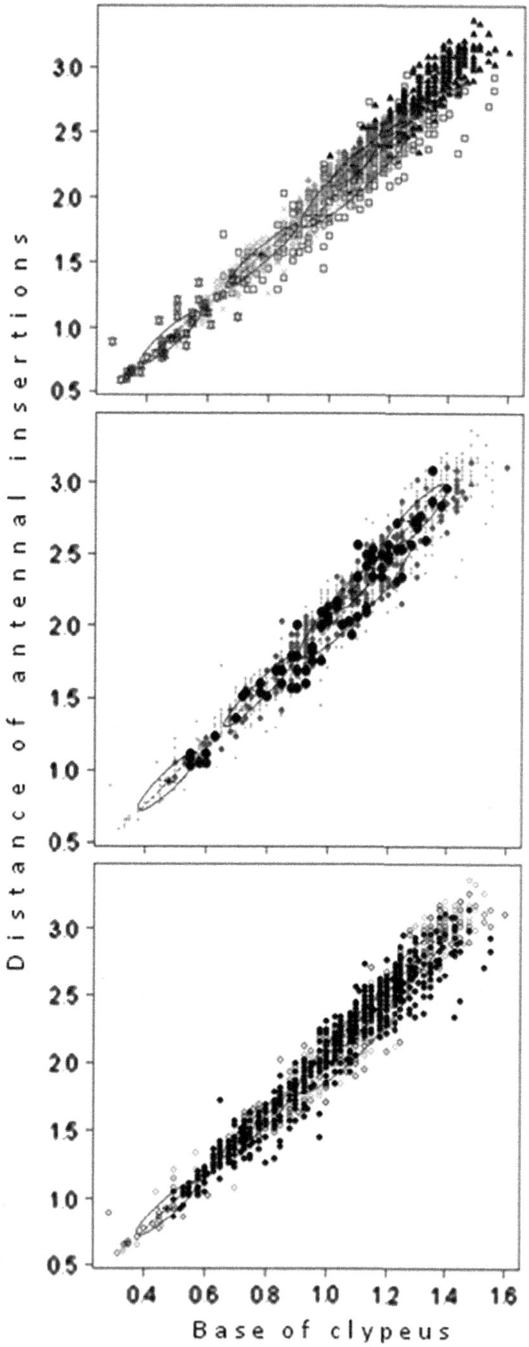

To remove the disagreement among variables we repeated the analysis but now the two principal components (PC1 and PC2) of the previous PCA analysis comprised the independent variable set. The same approach was taken by Hunt and Chapman (2001) to concentrate the entire variation in just one variable. The philosophy is substantiated by the mathematical arguments used by Fraley and Raftery (2007). The results of this analysis are shown in Fig. 10.16.

The number of clusters is not necessarily the number of instars when principal components are analyzed unless all components have the physical meaning of size but this is not the case in the study of *S. vestita* larval instars. In Fig. 10.16 it can be seen that the central cluster represented by a big circle includes all larvae deviating in shape (spiracle length/width). The same applies to the ellipses included in the circle, relating to larvae of L3 and L4 instars. The clusters L1–L5 are all located in parallel along the PC1 size axis and for this they are considered instar groups. In terms of BIC values the model that takes preponderance is the one that comprises

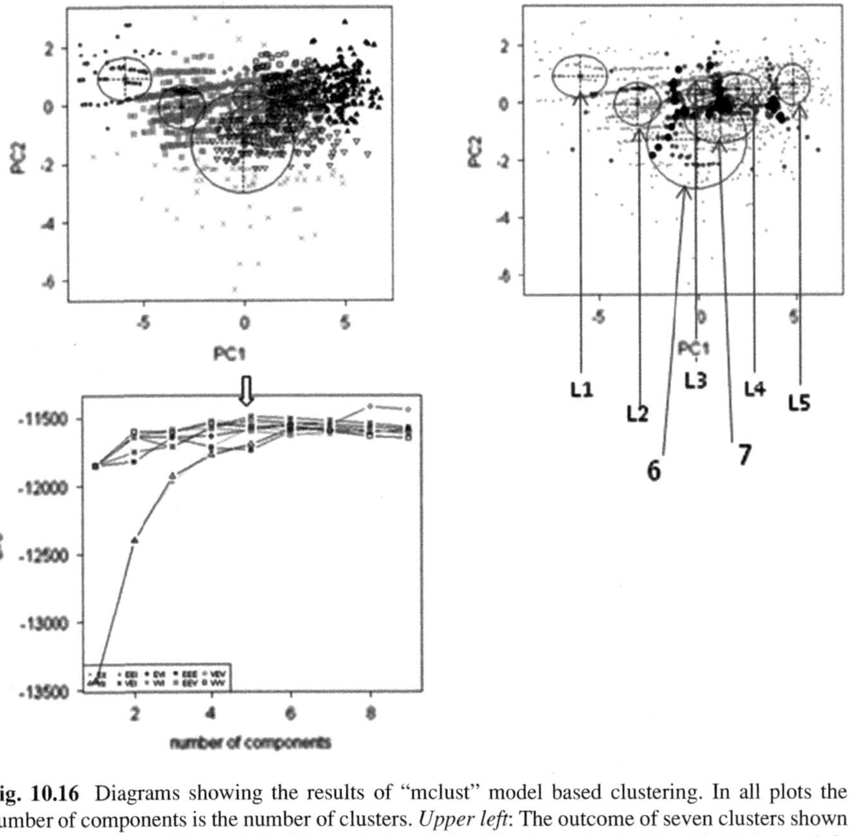

Fig. 10.16 Diagrams showing the results of "mclust" model based clustering. In all plots the number of components is the number of clusters. *Upper left*: The outcome of seven clusters shown on a PC1–PC2 plot. *Upper right*: Uncertainties of classification as in the Fig. 10.15. *Lower left*: BIC values for different models. The arrow labeled L1–L5 shows the best BIC value supported by the VVV and VEI models. The *arrows* 6 and 7 are distributed in L3, L4 in the five cluster best BIC solution (see text)

8, 9 or 5 clusters (Fig. 10.16 Lower left) and corresponds to the VVV[1] and VEI[2] models of Fraley and Raftery (2007). Seven clusters describe the meaning of the biologically decoupled clusters and the solution involving eight or nine clusters is entirely out of the biology of beetles. If these clusters were also plotted they would have crowded the plot making it impossible to be inspected by readers. In many studies the large number of clusters indicates outlying measurements and the excessive clusters are wasted by including them all. Therefore the five clusters version is the best model describing the morphometric data since it has the maximal BIC value under several models while being biologically sound.

On the other hand as Johnson and Williamson (2006) stated, the closely related cerambycids in the same subfamily (Lamiinae) have five larval instars while the more distant taxonomically *Phoracantha semipunctata* which belongs in the subfamily (Cerambycinae) has six instars. In effect five instars against six are preferred in their study.

The number of putative instars and the cluster affiliation of each individual larva is shown in Fig. 10.17 together with the number of larvae in each cluster. In natural populations as we proceed to more advanced instars there is a decrease in the number of individuals as a result of mortality factors unless the duration of each instar is different which does not seem in the case of Cerambycidae beetles. In wood feeding insects the change of feeding site from the cambial zone to heartwood causes an additional decrease in the number of larvae as noted by Johnson and Williamson (2006). This is not reflected in Fig. 10.17 where the L3 stage is the most numerous.

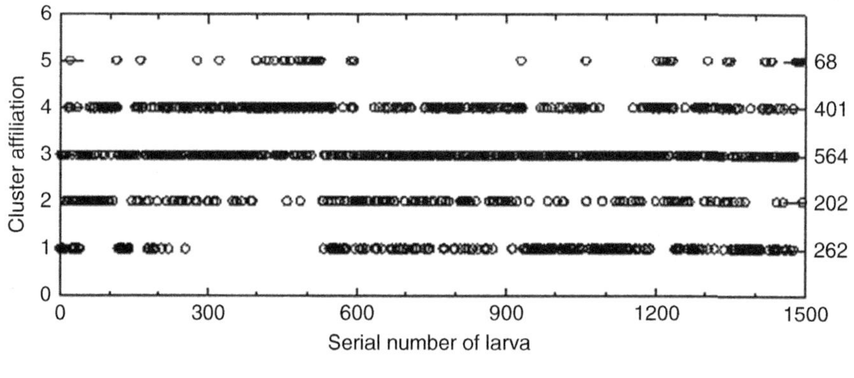

Fig. 10.17 Cluster affiliation of individual larvae on the basis of misclassification probabilities. The same larvae that produced the diagrams in Fig. 10.15 are used. The numbers at the right show the number of larvae in each instar cluster

[1]VVV: Variable volume and shape of components with variable orientation of ellipses representing clusters.

[2]VEI: Variable volume of components, which have the same shape and the same orientation in relation to coordinate axes.

Table 10.5 Correlations of measured variables with the probabilities of affiliation to individual clusters [1...5] entered as a categorical variable

Measured variable	Instar-1	Instar-2	Instar-3	Instar-4	Instar-5
A. Left labial palpomere length	0.328 (0)	−0.106 (ns)	−0.084 (ns)	−0.203 (0)	−0.556 (0)
B. First thoracic spiracle length	0.52 (0)	−0.041 (ns)	−0.156 (0.002)	−0.302 (0)	−0.826 (0)
C. First thoracic spiracle width	0.604 (0)	−0.308 (0)	−0.012 (ns)	−0.188 (0.002)	−0.629 (0)
BC. First thoracic spiracle length/ width[a]	−0.163 (ns)	0.566 (0)	−0.258 (0)	−0.229 (0)	−0.431 (0.003)
D. Base of labrum	0.53 (0)	−0.151 (ns)	−0.042 (ns)	−0.195 (0.001)	−0.675 (0)
E. Base of clypeus	0.556 (0)	−0.107 (ns)	−0.068 (ns)	−0.298 (0)	−0.647 (0)
F. Distance of antennal insertions	0.684 (0)	−0.125 (ns)	−0.114 (ns)	−0.278 (0)	−0.703 (0)
G. Head width	0.691 (0)	−0.111 (ns)	−0.091 (ns)	−0.246 (0)	−0.663 (0)
H. Maximal prothoracic width	0.639 (0)	−0.113 (ns)	−0.126 (0.03)	−0.244 (0)	−0.743 (0)
I. Maximal prothoracic width	0.607 (0)	−0.149 (ns)	−0.072 (ns)	−0.205 (0)	−0.727 (0)
J. Distance of labrum to prothorax	0.435 (0)	−0.067 (ns)	−0.14 (0.009)	−0.289 (0)	−0.473 (0.001)

The numbers in brackets are the associated Bonferroni probabilities of getting the respective correlation by chance alone; if smaller than 10^{-5} they are entered as 0
ns non significant correlations
[a]The line is painted in grey to indicate the differentiation of BC as shape variable

The numbers of larvae in clusters can be seen also in Fig. 10.16 where the central clusters 6 and 7 have been incorporated in L3, L4 and partly in L2 instars.

In the correlation of the measured morphometric variables, each instar exhibits its own pattern as in Johnson and Williamson (2006), and there is no consensus among the variables and the instar groups. Table 10.5 summarizes all correlations of the original variables measured on a larva with the maximum probability of affiliation to a cluster. The largest probability, instead the probabilities of affiliation to every cluster, were taken in order to lower the calculated correlations and produce comparable numbers. All the correlations apply only within each cluster.

From Table 10.5 is evident that there is no single variable that is strongly correlated with all clusters. This means that in Cerambycidae beetles there is no morphometric variable, among those measured, to account for the increase in size and the molts to instar groups (clusters). Moreover, there is no agreement between variables. The first instar seems to correlate well with all variables, except the shape variable BC, and the same is observed in the fifth instar though with minus sign. The intermediate instars 2, 3, and 4 are correlated to the shape variable BC and show low and non significant correlations to the size variables. The reader should recall that the configuration of larvae was done on all variables through their principal scores. In Lepidoptera this problem disappeared since only the length of the epicranial sclerites of larvae was measured (Petrakis 2000). In the paleontological study of the trilobite *A. bellatula* the two measured variables were reduced to one by taking the principal score (Hunt and Chapman 2001). In general there is tendency to reduce the number of variables that determine the instar affiliation of larvae. Johnson and Williamson (2006) did not follow this tendency. Instead, they performed a reliability analysis and found that the variables F, G, E and I are the most reliable in producing the instar affiliation. In this analysis they determined a consensus instar as the most frequent instar found in all instar groupings from all individual variables. They then calculated the reliability of each variable as the percentage of instar matches between the consensus classification and the classification on this variable.

The idea behind the reliability analysis is followed in this paper. However, the geometric progression of larval characters with molts was not considered. Instead the variables were treated collectively in a PCA scheme and the principal scores were subjected to a model based classification while the model was a normal curve (one for each size class = putative instar). The reliability of each morphometric variable was then expressed as a correlation coefficient for each instar. This is impractical in instar recognition because the investigator has to know a priori the instar affiliation. The methods of model based clustering and the subsequent procedures are presented for the analysis of the effect and the relative importance of each variable in each cluster. Nevertheless this analysis, as that in of Hunt and Chapman (2001) presents a statistical test for the number of clusters by means of the best BIC number.

Ideally the increase in size among successive instars would be studied with morphometric measurements in the same individual larva. However, this is a very difficult problem since the insect lives in the bark and the wood of linden trees digging a feeding gallery which ends to a pupation site. Any type of disturbance exerted by the investigator on the feeding site would have unpredictable affects on the growth process of the insect.

Conclusions

The tools on the desk top of ecologists, taxonomists and paleontologists have increased greatly in the last 20 years. Simultaneously the cost of statistical and data acquisition software products and the associated machinery have decreased substantially or are distributed for free through the global web networks connecting computers.

The increase in methods, algorithms and instrumentation is addressed to a long standing biological question: how many larval instars can be detected in a given collection of morphometric data? The suggested methodology is exemplified by four examples (*P. oleae*, *C. sellata*, *A. bellatula*, and *S. vestita*). Three main computer packages were used for these case studies. The commercial SYSTAT 9 package (Wilkinson 1999), the free educational package PAST (Hammer et al.

2009) and the freely available R library (R Development Core Team 2008) with the libraries *mclust* (Fraley and Raftery 2007, 2009), *flexmix* (Gruen and Leisch 2008), *mixreg* (Turner 2009) and *rpart* and *mvpart* (Therneau and Atkinson 2008, 2009) were used in all analyses and presentations. The case of leaf generation (the feeding mode is a leaf-mining larva) of the trivoltine olive kernel moth *P. oleae*, in a set of six olive cultivars (Petrakis 2000) and varieties is the first example. The second example concerns the trilobite *A. bellatula* and the third example the trilobite *C. sellata* (Hunt and Chapman 2001). The fourth example *S. vestita* is problematic because many of larval morphometric variables produce different instar group numbers (Johnson and Williamson 2006). Discriminant analysis, hierarchical cluster analysis, classification binary trees (Breiman et al. 1984) and model-based clustering were employed. The important property of model-based clustering to statistically test the number of instar groups was in all cases a confirmation of the original suggestion of the researchers. The philosophy underpinning the methodology of model-based clustering facilitated the discussion of instar grouping and the effect of each morphological variable to the various larval instars. The inclusion of many morphometric variables in the case of *S. vestita* permitted the distinction between size and shape variables.

It is also expected that in the not too distant future there will appear new algorithms more biologically oriented on the model-based clustering. Together with the more rapid and accurate data acquisition methods they are expected to boost the power of researchers and throw new light to the growth of invertebrates.

Acknowledgements Many people contributed in this manuscript. Many ideas emanated from discussions with A. Feest (University of Bristol), K. Spanos (Forest Research Institute in Thessaloniki, Genetics) and A. Legakis (University of Athens, Biology). The data for trilobites and *S. vestita* were provided by G. Hunt (Smithsonian Institution, Paleobiology) and R.C. Williamson (University of Wisconsin, Entomology), V. Roussis and the late C. Vagias (University of Athens, Pharmacy), P. Dais (University of Crete, Chemistry) are thanked for the applications of some methods in chemical characters of insect and plant species. V. Petrakis (University of Athens, Archaeology) provided important literature and advice on past Mediterranean ecology and uses of organisms by humans. G.P. Patil (University of Pennsylvania, Mathematics) provided literature. D. Roumeliotis is thanked for the writing some utility programs and E. Lahlou helped in typing literature entries.

Appendix

See Figs. 10.18–10.20

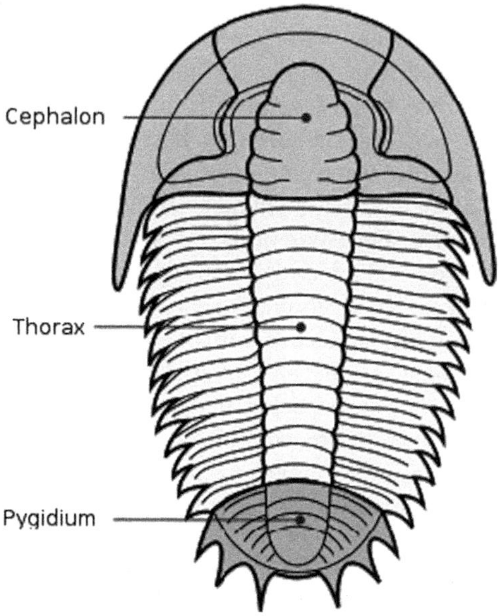

Fig. 10.18 Generalized shape of a trilobite body on which the cephalon length and width are measured (Hunt and Chapman 2001)

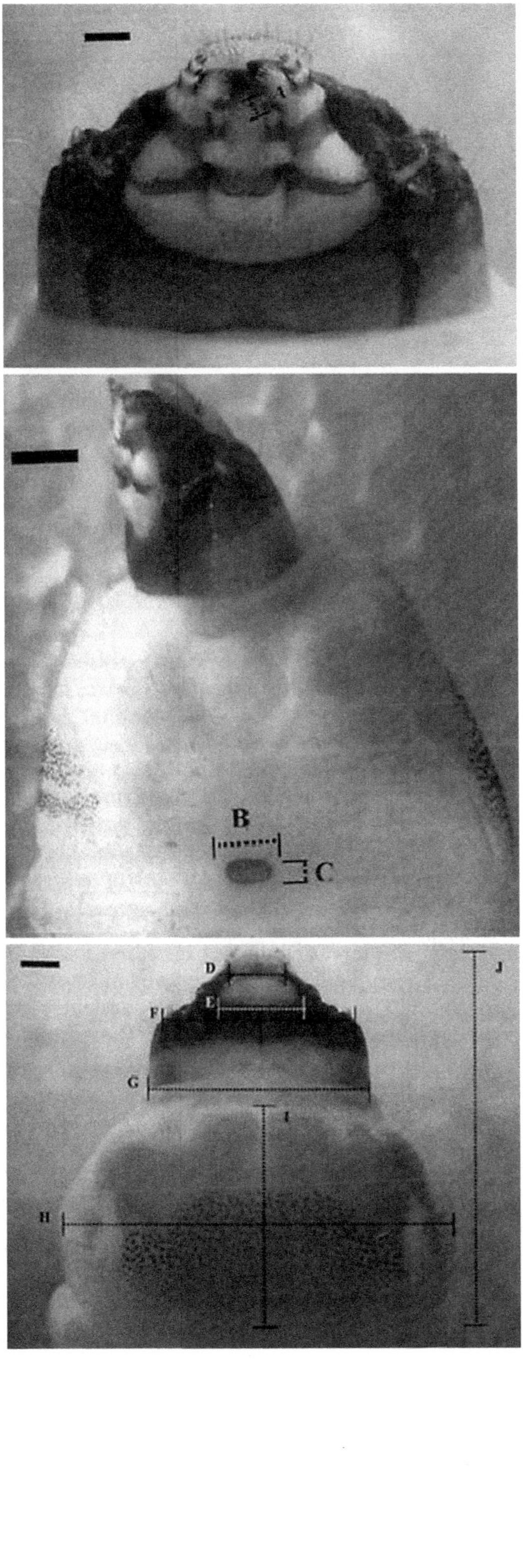

Fig. 10.19 Images of the fifth instar larval head and thorax body on which morphometric measurements were taken (after Johnson and Williamson 2006). The following measurements are as follows. A: Left palpomere length, B: First thoracic spiracle length, C: First thoracic spiracle width, D: Clypeolabral suture length, E: Frontoclypeal suture length, F, G: Cranial widths, H: Maximal prothoracic width, I: Mesal pronotal length, J: Length of labrum to prothorax

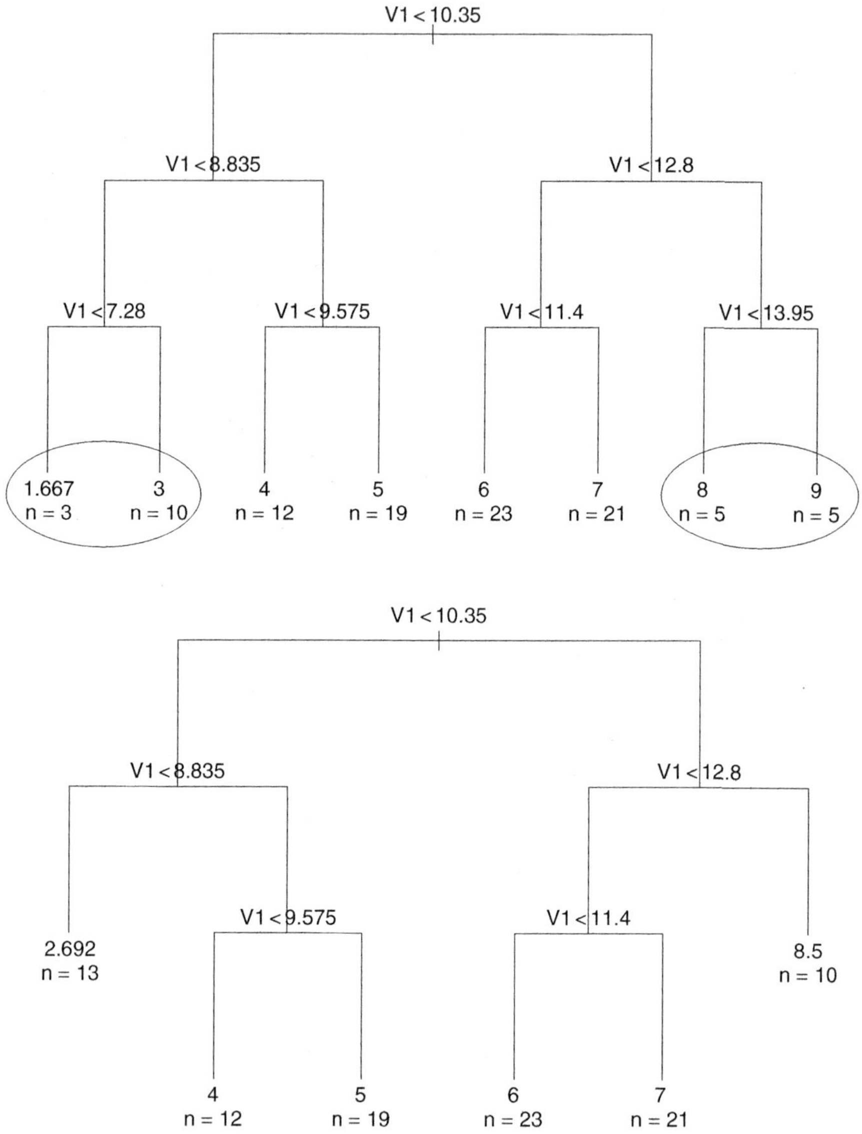

Fig. 10.20 Two dendrograms of the *A. bellatula* data on all 98 individuals produced by employing the R library *rpart* (Therneau and Atkinson 2008). The upper dendrogram has eight and the lower six terminal leaves (=instar groups). Splitting values are very close to the ones in Fig. 10.9. The two ellipses contain the groups that collapse into the groups in the lower plot. The variable V1 denotes the cephalon length. The cephalon width (=V2) is used as a surrogate variable that in all terminal leaves agrees with V1 and is considered in the calculation of splits

References

Alvan-Aguilar M A & Hamada N 2003 Larval Biometry of *Simulium rubrithorax* (Diptera: Simuliidae) and Size Comparison between Populations in the States of Minas Gerais and Roraima, Brazil Mem Inst Oswaldo Ctuz, Rio de Janeiro 98:507–511

Appleby JH and Credland PF (2007) The role of temperature and larval crowding in morph determination in a tropical beetle, *Callosobruchus subinnotatus*. Journal of Insect Physiology 53: 983–993

Beaver RJ and Sanderson JP (1989) Classifying instars of the navel orangeworm (Lepidoptera: Pyralidae) based on observed head capsule widths. Journal of Economic Entomology 82: 716–720

Benaglia T, Chauveau D, Hunter DR, Young D (2009) mixtools: An R Package for Analyzing Finite Mixture Models. Journal of Statistical Software 32:1–29

Breiman L, Friedman J, Olshen R, Stone, C (1984) Classification Regression Trees, Wadsworth, California, USA,.

Brezinski DK (1986) An opportunistic Upper Ordovician trilobite assemblage from Missouri. Lethaia 19:315-325

Bybee SM and Branham MA (2008) Scanning Rocks for Data. American Entomologist 54: 214–217

Chang ES (1993) Comparative endocrinology of molting and reproduction: Insects and crustaceans. Annual Review of Entomology 38: 161–180

Chatterton BDE, Edgecombe GD, Speyer SE, Hunt AS, Fortey RA (1994) Ontogeny and relationships of Trinucleoidea. Journal of Paleontology 68:523–540

Cisne JL (1973) Life history of an Ordovician trilobite *Triarchus eatoni*. Ecology 54:135–142

Crosby TK (1973) Dyar's Rule predated by Brooks' Rule. N.Z. Entomology 5:175–176

Daly HV (1985) Insect Morphometrics. Annual Review of Entomology 30:415–38

De'ath G and Fabricius KE (2000) Classification and Regression Trees: A powerful yet simple technique for ecological data analysis. Ecology 81: 3178–3196.

Dyar HG (1890) The number of molts of lepidopterous larvae. Psyche 5:420–422

Elewa AMT (2007) A Powerful Electronc Journal in the New Millenium. Paleontologica Electronica, 10: 1–2

Elewa AMT (2009) The role of paleontology in the "Electronic Era". Journal of Geology and Mining Research 1(10): 1–2

Elewa AMT (2010) Paleoinformatics: The superhighway to modern paleontology. Journal of Geology and Mining Research 2(1): 1–2

Elmore RT, Hettmansperger TP, Thomas H (2004) Estimating component cumulative distribution functions in Finite Mixture Models." Communications in Statistics - Theory and Methods 33:2075–2086

Engelman L (1999) Discriminant Analysis SYSTAT9: Statistics I. Wilkinson, L. [ed] SPSS Inc., Chicago, USA

Floater GJ (1996) Life history of ground- and canopy-nesting populations of *Ochrogaster lunifer* Herricj-Sheffer (Lepidoptera: Thaumetopoeidae): evidence for two species? Australian Journal of Entomology 35: 223–230

Fraley C, Raftery AE (1998) How many Clusters? Which Clustering Method? Answers Via Model-Based Cluster Analysis. University of Seattle, Dept. of Statistics, Technical Report 329

Fraley C and Raftery A (2007) mclust Version 3 for R: Normal Mixture Modeling and Model-Based Clustering. University of Seattle, Dept of Statistics, Technical Report 504.

Fraley C and Raftery A (2009) mclust: Model-Based Clustering and Normal Mixture Modeling. R package version 3.3.1 URL http://CRAN.R-project.org/package=mclust

Gelman DB, Kelly TJ, Reed DA, Beckage NE (1999) Synthesis/release of ecdysteroids by *Cotesia congregata*, a parasitoid wasp of the tobacco hornworm, *Manduca sexta*. Archives of Insect Biochemistry and Physiology 41: 17–29

Goldson SL, McNeill MR, Proffitt JR & Baird DB (2001) Seasonal variation in larval-instar head capsule sizes of Argentine stem weevil, *Listronotus bonariensis* (Kuschel) (Coleoptera:Curculionidae). Australian Journal of Entomology 40:371–375

Gruen B and Leisch F (2008) FlexMix Version 2: Finite Mixtures with Concomitant Variables and Varying and Constant Parameters. Journal of Statistical Software 28:1–35

Hammer O and Harper D (2006) Introduction Paleontological Data Analysis. Blackwell Publishing

Hammer O (2004) Allometric field decomposition: An attempt at morphogenetic morphometrics. In "Morphometrics - Applications in Biology and Paleontology" Elewa AMT (ed) Springer Verlag, Berlin

Hammer O, Harper DAT, Ryan PD (2009) PAST - Paleontological Statistics.ver. 1.90 URL http://folk.uio.no/ohammer/past

Hanks LM and Denno RF (1993) The role of demic adaptation in colonization and spread of scale insect populations in : Evolution of Insect Pests - Patterns of variation Kim K C and McPheron B A (ed), New York, Wiley

Hopkins TL, Kramer KJ (1992) Insect cuticle sclerotization. Annual Review of Entomology 37: 273–302

Hunt G and Chapman RE (2001) Evaluating hypotheses of instar-grouping in arthropods: a maximum likelihood approach. Paleobiology 27: 466–484

Hunt G (2001) A Quick Guide to Mixture Model Analysis version 1.31, manual. Committee on Evolutionary Biology, University of Chicago

Hutchinson GE and Tongring N (1984) A possible adaptive significance of the Brooks-Dyar rule. Journal of Theoretical Biology 106:437–439

Hutchinson JMC, McNamara JM, Houston A, Vollrath F (1997) Dyar's Rule and the Investment Principle: optimal moulting strategies if feeding rate is size-dependent and growth is discontinuous. Philosophical Transactions of the Royal Society of London B352:113–138

Jiao WJ, Ling Z, Wen LG, Qiu XW (1998) The leafmining behavior and instars of *Liriomyza sativae* (Diptera: Agromyzidae). Journal of the South China Agricultural University 19: 27–31

Johnson RA and Wichern DW (1998) Applied Multivariate Statistical Analysis. Prentice Hall, New Jersey, USA

Johnson TA and Williamson RS (2006) Multiple morphological measurements as larval indicators for *Saperda vestita* (Coleoptera: Cerambycidae). Annals of the Entomological Society of America 99: 938–44

Jolicoeur P (1999) Introduction to Biometry. Kluwer Academic, New York

Kishi Y (1971) Reconsideration of the methods to measure larval instars by use of the frequency distribution of head-capsule widths or lengths. Canadian Entomologist 103: 1011–1015

Kopaska-Merkel D (1981) Trace-fossil frequency modes and arthropod growth. Northeastern Geology 10:300–306

Legendre L and Legendre P (1983) Numerical Ecology. Elsevier, Amsterdam

Lewandowski M, Sznyk A & Bednarek A (2004) Biology and morphometry of *Lycoriella ingenua* (Diptera: Sciaridae). Biological Letters 41: 41–50

Lubischew AA (1962) On the use of discriminant functions in taxonomy. Biometrics 18: 455–477

Morrison DF (1984) Multivariate Statistical Methods. McGraw-Hill, Singapore

Mosimann JE and Malley JD (1979) Size and shape variables, In "Multivariate Methods in Ecological Work", (eds) L Orloci, CR Rao, WM Stiteler, ICPH, Burtonsville, Md, USA

Oksanen J, Kindt R, Legendre P, O'Hara R, Simpson GL, Stevens MHH, and Wagner H (2008) vegan: Community Ecology Package. R package version 1.13-2. http://vegan.r-forge.r-project.org/ (accessed 17.05.2008)

Orloci L and Kenkel NC (1985) Introduction to Data Analysis with Examples from Population and Community Ecology. ICPH, Burtonsville, Md

Page M, Nelson LJ, Haverty MI & Blomquist GJ (1990) Cuticular hydrocarbons as chemotaxonomic characters for bark beetles: *Dendroctonus ponderosae, D. jeffreyi, D. brvicois,* and *D. frontalis* (Coleoptera: Scolytidae). Annals of the Entomological Society of America 83, 892–901

Page M, Nelson LJ, Forschler BT, Haverty MI (2002) Cuticular hydrocarbons suggest three lineages in *Reticulitermes* (Isoptera: Rhinotermitidae) from North America. Comparative Biochemistry and Physiology 131B: 305–324

Petrakis PV (2000) Larval performance in relation to oviposition site preference in olive kernel moth (*Prays oleae* BERN., Yponomeutidae, Praydina). Agricultural and Forest Entomology 2: 271–282

Petrakis PV, Agiomyrgianaki A, Christophoridou S, Spyros A, Dais P (2008) Geographical characterization of greek virgin olive oils (Koroneiki cv) using 1H and 31P NMR fingerprinting with canonical discriminant analysis and classification binary trees. Journal of Agricultural and Food Chemistry 56: 3200–3207

Pimentel RA (1992) An introduction to ordination, principal component analysis and discriminant analysis. In 'Ordination in the Study of Morphology, Evolution and Systematics of Insects', Sorensen, J. T. & Foottit, R. [eds], Elsevier, Amsterdam

R Development Core Team, (2008). R: A language and environment for statistical computing. R Foundation for Statistical Computing, Vienna, Austria.

Scott EL (1979) Correlation and suggestions of causality: spurious correlation. In "Multivariate Methods in Ecological Work", (eds) L Orloci, CR Rao, WM Stiteler, ICPH, Burtonsville, Md

Simpson GL (2007) Analogue methods in palaeoecology: using the analogue package. Journal of Statistical Software 22: 1–29

Soponis AR and Russell C (1982) Identification of instars and species in some larval *Polypedilum* (*Polypedilum*) (Diptera: Chironomidae). Hydrobiologia 94:25–32

Steinberg D and Colla P (1997) CART–Classification and Regression Trees. Salford Systems; San Diego, CA.

Taylor PC and Silverman BW (1993) Block diagrams and splitting criteria for classification trees. Statistics and Computing 3: 147–161

Therneau TM and Atkinson B (2008) rpart: Recursive partitioning. R package version 3.1–41, R port by Brian Ripley. URL http://CRAN.R-project.org/package=rpart

Therneau TM and Atkinson B (2009) mvpart: Multivariate regression trees. R package version 1.2–6, R port by Brian Ripley, extensions and adaptations of rpart to mvpart by G. De'ath. URL http://CRAN.R-project.org/package=mvpart

Turner R (2009) Functions to fit mixtures of regressions. R package version 3.3.1 URL http://CRAN.R-project.org/package=mclust

Wilkinson L (1985) Mobiles. Northwestern Univ., Dept of Statistics, Evanston, Illinois

Wilkinson L (1999) SYSTAT 9: Statistics I & II Evanston, Illinois

Whittington HB (1990) Articulation and exuviation in Cambrian trilobites. Philosophical Transactions of the Royal Society of London (B) 324: 111–47

Yuan LL (2007) Maximum likelihood method for predicting environmental conditions from assemblage composition: The R Package bio.infer. Journal of Statistical Software 22: 1–20

Chapter 11
Future Insights in Computational Paleontology: With Special Spotlight on Visual Paleontology

Ashraf M.T. Elewa

The University of California Museum of Paleontology (UCMP) divided paleontology into the following traditional subdisciplines: micropaleontology, paleobotany, palynology, invertebrate paleontology, vertebrate paleontology, human paleontology (paleoanthropology), taphonomy, ichnology, and paleoecology.

However, I would add the following modern trends (all can be attributed to computational paleontology): Paleoinformatics (paleontological information Systems), molecular paleontology, and may be visual paleontology (could be a new term for old branch).

Molecular paleontology may be founded in 1956 by Abelson, who made the first attempt to recover the proteinaceous components of fossils. Since that time, this field has received several attempts through new techniques that enabled paleontologists to apply new analytical methods to fossils.

Relating to paleoinformatics, we urgently need, in my opinion, to prepare international taxonomic databases for paleontologists and biologists (especially taxonomists) (see Elewa 2010). It is notable that several modern organisms that have resemblances in the past are attributed to different taxonomic identities!!

The story is distinctly different for visual paleontology; this branch needs laboratories prepared with expensive facilities for modeling and visualization. Nonetheless, many colleagues could widen our knowledge through publishing interesting research on the subject.

In 2002, Allen Debus and colleagues published a book titled "Paleoimagery: The Evolution of Dinosoaurs in Art". They explained how paleoartists can depict scientific ideas about dinosaurs and prehistoric animals on canvas and in sculpture. This subject is one of the most important themes in visual paleontology.

A.M.T. Elewa
Geology Department, Faculty of Science, Minia University, El-Minia, Egypt
e-mail: aelewa@link.net, ashrafelewa@ymail.com

A.M.T. Elewa (ed.), *Computational Paleontology*,
DOI 10.1007/978-3-642-16271-8_11,

Karl Bates et al. (2008) used laser scanning (LiDAR), which provides sufficient resolution to perform robust quantitative analysis of dinosaur tracks, and computer modelling methods to create a range of 3D models of the Fumanya dinosaurs specimens, attempting to reconstruct their body sizes and shape as in life.

Karl Bates et al. (2009) stated that using laser scanners and digitization methods, complex concepts (including paleontological concepts) can be shown to members of the public in a highly visual manner. They explained how to use scan of a tracksite to highlight tracks that would otherwise be hard to see.

Peter Falkingham and his colleagues (2009) used Finite Element Analysis (FEA) to give new insights on the evolution of webbing in early birds, as well as reported webbing in other vertebrates.

One of the most interesting subjects is that was published by SandiaLabNews in December 19, 1997 (http://www.sandia.gov/LabNews/LN12-19-97/dinosaur_story.html). Scientists and paleontologists from Sandia Laboratories and the New Mexico Museum of Natural History collaborated together to take computer analysis into new realms to recreate voice from the Late Cretaceous Parasaurolophus dinosaur.

"A History of Paleontology Illustration" is a book was published by the Indiana University Press (2008). The author of the book (Jane Davidson) presented a historical review of paleontological illustrations since the fifteenth century to the beginning of the twentieth century.

In her article on science and art, Marry Parrish (2008) declared that the book of Davidson offers interesting and well-constructed overviews of the natural history and imagery of birds and fossils. Brian Switek, in his review to the same book, affirmed that the techniques used to bring ancient creatures back to life provides important details in each section, and there can be little doubt that the efforts of artists have always been indispensable to paleontologists (ScienceBlogs.com: http://scienceblogs.com/laelaps/2008/08/book_review_a_history_of_paleo.php). In another review, Switek (2009) argued that Davidson clearly drew on a wide variety of sources, from Renaissance paintings to illustrations in scientific catalogs, and her book is a unique and important contribution to the history of paleontology.

Of course, there are several subjects of interest that are not covered herein, nevertheless the current review is aimed at impelling paleontologists to open new windows on modern technologies to ensure the continuity and applicability of paleontology in the scientific community, from one side, and from the other side to simply facilitate complex concepts related to this science to the public.

In summary, either we (paleontologists) concentrate our efforts to develop new and modern trends in the field, in which we will be able to attract more audience from our communities, or paleontology will be, unfortunately, considered as one of the "old fashion" sciences.

Finally, I would mention the words of Jere Lipps (2000), who said "Paleontology requires creativity. Why? Because it is not an easy science. It is a way to understand the history of life through repeated, reliable observations as well as hypothesis development and testing in the face of limited and often confusing data".

References

Abelson P (1956) Paleobiochemistry. Scientific American, 195:83–92

Bates K, Rarity F, Manning P, Hodgetts D, Vila B, Oms O, Galobart A, Gawthorpe R (2008) High-resolution LiDAR and photogrammetric survey of the Fumanya dinosaur tracksites (Catalonia): implications for the conservation and interpretation of geological heritage sites. J Geol Soc 165 (1); 115–127

Bates K, Falkingham P, Hodgetts D, Farlow J, Breithaupt H, O'Brien M, Matthews N, Sellers W, Manning P (2009) Digital imaging and public engangement in palaeontology. Geology Today 25 (3): 95–100

Davidson J (2008) A History of Paleontology Illustration. Indiana University Press, Bloomington, 235 pp

Debus A, Debus D, Glut D (2002) Paleoimagery: The Evolution of Dinosoaurs in Art. McFarland & Company, Incorporated Publishers 293 pp

Elewa AMT (2010) Paleoinformatics: The Superhighway to Modern Paleontology. Journal of Geology and Mining Research, Academicjournals 2(1): 2 pp

Falkingham P, Margetts L, Smith I, Manning P (2009) Reinterpretation of palmate and semi-palmate (webbed) fossil tracks; insights from finite element modelling. Palaeogeography, Palaeoclimatology, Palaeoecology 271 (1-2): 69–76

Lipps JH (2000) Creative paleontology. Palaeontologia Electronica 3 (2): 4 pp

Parrish M (2008) Science and Art: Subject Matter Matters. Science 322 (5899): 196–197

Switek B (2009) Review of the book titled "A History of Paleontology illustration". Palaeontologia Electronica 12 (1): 3 pp

Index

A.M.T. Elewa (ed.), *Computational Paleontology*,
DOI 10.1007/978-3-642-16271-8, © Springer-Verlag Berlin Heidelberg 2011

9783642162701